9/18-EBR-ZERO

To order additional copies, please contact us.
BookSurge, LLC
www.booksurge.com
1-866-308-6235
orders@booksurge.com

9/18-EBR-ZERO
Hauntings of Katrina

Andree' M. Maduell

2006

9/18-EBR-ZERO

IN MEMORY OF RICKY

This book is dedicated to the rescue workers, military and civilian, the reporters who kept us informed, and the volunteers who gave their time, and the thousands of ordinary people who showed concern with donations.

PROLOGUE

We were once a great city. We were born out of the waters of the great Mississippi River. We became the Heart that pumped life into America. We grew into the "City That Care Forgot," but we forgot to care about what was important. We forgot to care about our sick and needy. We forgot to care about our children and our old. We became complacent as we went about our daily lives thinking that nothing would change, no one would bother us, and life would go on as usual.

Nature had other plans for us. Nature hit us over the head with a blow that was felt around the world. Suddenly, we became the main story, the focus of attention to all who knew us, and those who wish they had. Party City drowned. The world was stunned. Soon everyone began to have their opinions about why we were hit. "They sinned," they said. "They were stupid to live below sea level." "They were cheated on by greedy and/or inept planners." "They did it to themselves." "Politicians did it." "God did it." "The devil did it."

It was nobody's fault, it just happened.

Now we have major decisions to make. What do we do now? Who should make the decisions? Should we leave or rebuild? Should we build it like it was, or should we become a plastic copy of a comic book future? Or is there another way? No answer is right or wrong. No answer will please everyone.

Ideas for our future are coming up like daisies in the springtime. The newspaper is full of opinions, articles, and columns from the serious writers to the silly. Many of these ideas are serious, some are in jest, and some are just plain flippant. No matter what these ideas are, I think that they should ALL be looked at and considered. Wasn't it a crazy idea that the earth revolved around the sun instead of the other way around? Wasn't it outrageous that the earth was round and not flat? People were ostracized, imprisoned, and even put to death for such ridiculous ideas. Why should we NOT listen to even the craziest ideas? Could it be because they cost too much? Could it be that the politicians refuse to give up control of their areas and the finances that they have control of? Politicians are using emotion rather than logic to get the people back. They keep reporting that it is safe and that people are returning by the droves, even though that is not true. Politicians want their tax base to return in order to keep their jobs. They know that the people want to return on an emotional level. They skirt around the issue of "Is it safe?" They quote "experts" who report that it is safe, and ignore those with opposite opinions. We need to make our own decisions based on our own reasons and logic. We can decide for ourselves what our future should be like, and the future of our cities.

1.
Random Acts of Nature

I did exactly what they said NOT to do. I stayed. I looked out my windows. I went outside. It was impossible NOT to watch as Nature chose my little corner of the world to unleash Her full power on. I knew that it was probably not a wise thing to do, but I couldn't help myself. I struggled over my decision to stay or go, but my curiosity won over my responsibility. The excitement of witnessing Hurricane Katrina first hand was too much to ignore. So I decided not to evacuate, but instead ride it out in my new home. I had recently moved from Metairie to Lacombe, a quiet peaceful place north Lake Pontchartrain. I felt that I would be safe there because it was a little farther north of the Gulf of Mexico, and the hurricane might lose some of its intensity by the time it got to me, but I was wrong, very wrong. My new home is on seven acres of land with dozens of tall pine trees all around the house. Since the house is three stories I figured it would be a safe place to stay. If trees were to fall on the house, or if the water would rise, there would be some part of the house safe from damage. Truth was, I really did not want to leave my new house or my possessions, and I ignored what the cost might be. I felt invincible. I had waited too long and planned so much for my retirement days to be spent doing what I loved the most, painting and writing, that I could not believe that this random act of nature was going to destroy me or my dreams.

That Sunday evening, my oldest daughter, Cherie, decided to come over to stay with me. Cherie lived in a mobile home, thirty miles North of New Orleans. I thought it would not be safe for her there alone, so I asked her to come stay with me. When she arrived, I asked whether she wanted to stay or evacuate, secretly hoping that she would choose to stay, but she said that she would do whatever I decided. To complicate matters, she had Lupus and was a dialysis patient, and needed a treatment every four days. Since she had just had a treatment Saturday, I knew that she could go for about four days at the most, before needing another treatment. I was faced with having to change my mind and take her someplace safe, but where would that be? Would she be safer if we were stuck in traffic in the middle of the storm, or at my house with the comfort of shelter and a warm bed? Whatever I chose to do, I had to decide quickly because the window of safety for evacuating was rapidly closing. This time my decision was based on comfort believing that Hurricane Katrina would be no big deal, pass over swiftly with some tree or roof damage, and within a day or two everything would be back to normal. I had a false sense of security since the other hurricanes over the past few years caused minor damage. I figured that this one would be less than the officials had predicted, like always. I thought that the weather people and the news media liked to over dramatize those kinds of situations, perhaps to scare people into protecting themselves, and this would be another "cry wolf" occurrence. I remembered the other evacuations when people were stuck in their cars for hours on the highways trying to get out of the city. I had never left before and wasn't going to waste my time sitting in crawling traffic with wind and rain, breakdowns and tempers. After all, the other hurricanes were never as bad as expected, and I was sure this would be the same. So Cherie and I stayed in Lacombe.

The last time that Cherie and I were together during a powerful hurricane was forty years earlier for Hurricane Betsy. I was pregnant with her then and she was due any day. I stayed at my parents' house in Lakeview where, ironically, the worst damage was predicted to be. Across the street from the house was a levee, and the officials thought it might flood. Everyone was concerned, but in those days, no one recommended evacuating like today. All I wanted to do then was rest, but everyone kept checking on me to ask if I was all right, or did I think I wanted go to the hospital. I told them no, that I was fine, had no pains, and would feel better if they would let me sleep. Betsy was over quickly and caused no significant damage to my parents' house, or the neighborhood. Electricity was restored quickly, but other parts of the city were devastated by wind and water. I was not concerned about life outside my little world then, as none of the damage affected me. Cherie was born two weeks later with no problems. I considered naming her Elizabeth and calling her Betsy, but changed my mind. I'm glad I did.

Sunday morning was beautiful in Lacombe. The sky was bright, sunny, and clear. As the day wore on, a breeze began to blow and clouds started rolling in. By Sunday night, the wind started picking up and the rain came down harder with each passing hour. My electricity had already gone off before there was any wind or rain which annoyed me. I called Cleco, the local energy company, and quickly got a real person, not a recording, who told me that they were re-routing electricity to the New Orleans area. Well that was dumb, I thought. Why would they do that? Are we on the north shore less important than New Orleans?

There was nothing I could do about it. I figured I would get my electricity back in a day or two, so I was not that concerned. As the winds blew stronger, the rains increased and darkness fell. I lit candles in various rooms of the house. It wasn't that I had the foresight to stock up on candles, bottled water and snack food, which is the common ritual of most of us in the South as soon as a storm enters the Gulf. I had never prepared for a storm before. I had already been buying aromatic candles and drinking bottled water. I had a citronella candle in a pail that I lit to have some light outside the kitchen window. I put it under the carport attached to the house, where I intended to sit and watch the storm. I lit a couple of other candles throughout the house and was enjoying their pleasant aroma. I had two bedrooms downstairs for us to sleep in, but I knew I would not be going to sleep. For most of the "larger" hurricanes, I usually stayed up all night to watch. So I figured I was ready for this one.

As it got darker, the wind blew harder and the rains came down intermittently. The sound of both became increasingly louder. Watching became more and more fascinating. The intensity of it all was exciting. I went from window to window to see what was happening on my property. The noise became so loud that it sounded like a jet engine. I thought about the descriptions of tornadoes when people said that it sounded like a freight train, but this was constant and deafening, lasting through the night. As I watched, I could see groups of pine trees bending together, some eventually cracking, snapping, and crashing down from the strain of the 150-180 mph winds. I kept looking in the direction the wind was coming from to check on the condition of the trees that could possibly fall on my house. They were being blown wildly, but did not look like they would break. The pines close to the house were tall, but thin and appeared reasonably safe. Although the city was in darkness, I could see the black shapes of the trees dancing against the charcoal sky. At times, I would go outside and sit under the carport to watch and be in the midst of it all. The strength of the wind gusts pushed the door to the carport against me as if it did not want me to go out there, but I was able to get it open. Cherie and I sat close to the house but the wind did not seem that strong. Perhaps it was because we were on ground level under the protection of the carport. Perhaps the tall pines buffered us from the strongest of the wind's power. Amazingly, the mosquito candle outside blew out only once.

The flickering orange glow from the candle reflected on the white paint of my car and the beams of the carport, giving off as much light as a dim lamp. I felt an unbelievable eeriness as I sat there watching and listening, but it was not a frightening feeling. It was almost peaceful. All night we watched from one vantage point to another, moving around the house trying to find the best place to view the hurricane's fury. What was it that was so mesmerizing? I had been through hurricanes before, but this one was special. They were saying on the radio that this one was more dangerous than any other. This one was a category 5, larger than any I had ever been through before. This one was the "Big One." I wondered if I had made the right decision to stay, but it was too late to do anything now. Whatever would happen to Cherie and I would be on my shoulders now, and mine alone.

We would listen to the battery-powered radio at random intervals on 870 AM, the emergency station. I wanted to get an up to date, minute by minute account, but the reporters were becoming annoying with their reports of how high the winds were, or that it was too late to evacuate. They weren't telling me anything I didn't already know. What I wanted were the coordinates. At 9:30 P.M., they reported Katrina was 130 miles due south of New Orleans, the winds were 160 mph, and the coordinates were 27.2 N, and 89.1 W. New Orleans was exactly at 30/90. It was so close, but moving so slowly that it was going to be a very long night. When the reporters would talk of the devastation the storm could cause, (which they did constantly), I thought that it was only devastating because Man is in the way. This was Nature doing what Nature does, and we were powerless to do anything about it. We cannot stop Nature, or change Her, we can only watch and try to get out of the way. Sometimes the reporters would say that this building was destroyed or that area was flooded, but until there was proof, I listened with some skepticism since they had a tendency to exaggerate. Sometimes I turned off the radio for a while, partly to save what battery power was left in case I needed it later, and partly to rest my ears from Man's voice in order to hear Nature's voice. Besides, it was much more exciting to be experiencing this storm for myself.

At one point, as I watched from the third floor attic window, I was so enthralled that all I could think of was that if my life would end in this storm, at least it would be during one of the strongest moments from Nature. I wondered if anyone else felt that way. I wondered about other extreme forces of Nature: blizzards, tsunamis, earthquakes, volcanoes, tornadoes, if people who experienced them felt the way I did, with an overwhelming sensation of awe and helplessness! I don't remember ever feeling afraid. I don't understand why. Instead I felt calm and peaceful and even some joy at the glory of it all. There I was in the midst of one of the most powerful acts of Nature and I was not afraid.

I know that I was worried. Not for myself, but for my family. My sister, Rita, was going to stay also, but was talked into evacuating by her ex-husband who said that it was her responsibility to take care of her daughter and our mother. He said that I was foolish not to go too. Rita lived in Mandeville, about ten miles from me. My mother was in the process of moving from her house to my new house, and my niece, Alyssa, was a student at UNO.

My other niece, Lauren, had just moved to France with her seven-year-old daughter, Kaylee, so they were safely out of the way. Rita took Alyssa, Mama, their oldest cat, the hard drive to the computer, and a few items of clothing, and headed north on I-59 towards Jackson, Mississippi. We were still able to use our cell phones and called each other several times to check on each other. One of the last times we were able to call each other, she sounded upset.

"Where are you ?" I asked.

"We were trying to get to Jackson, but there are no rooms anywhere. I'm tired! I can't go anymore! I don't know what to do!"

"Didn't you say your friend, Linda, said you could go there?" I asked.

"It's her sister's house, and that's back in Alexandria."

"Well, go there!"

"Do me a favor first. Call motels and hotels and see if you can get us a room somewhere. Anywhere!"

I tried calling some motel 800 numbers and got what I thought I would get-no vacancies anywhere nearby. I called her back and told her, "No one has any rooms anywhere. Go to Linda's!"

"Oh, God! I'm so tired, I hope I can make it there. I'm turning off the Interstate now and will go back to Alexandria. I don't know what else to do!"

"Call me when you get there, I don't care what time it is, call me!"

I had also gotten a call from my youngest daughter, Nicolle. I knew that she and her family decided to evacuate, but I didn't know where they were going until she called me.

"Where are you?" I asked.

"We are on highway 90, on the Rigolets." (rig' o lees, the old highway that follows I-10 east to Florida.) "We booked a room on-line in Pensacola that takes dogs."

She was with her husband, Charles, their two children Charles and Nicolas, their dog, their in-laws, and another dog.

"The water is already over the road. We're looking for a gas station, but they are all closed. We filled up when we left, but the traffic is so bad, we are crawling and are almost on empty."

"Are you all in one car?" I asked.

"No. We took two cars. We didn't go on I-10 because we thought there would be less traffic here, but this is a nightmare!"

She called me once more when they got to the motel in Pensacola, ten hours later.

"There was a gas station just about to close, but they let a few more people in. We are at the motel now, and they are saying on the TV that the eye is just passing over St. Bernard Parish. You are going to be just west of the storm. Are you going to be ok?"

"We are fine now, but have no electricity. Call later and let us know what's happening."

There were no more phone calls. Communication was wiped out all over the entire Gulf South. I would not find out until days later the ordeal that my family later experienced.

Early Monday morning, the storm was still at full power. We tried to write down the coordinates as they announced them on the radio, which wasn't very often. We had a road atlas to check out where the eye was landing. It seemed as though it was going to go on forever. As fascinated as I was by it all, I was ready for it to be over. Just when it seemed to die down a bit, the gusts would come back as strong as before. Just when I thought we had come through it without any major damage, I noticed a trickle of water inside the house. A small puddle seemed to appear right in the middle of the house. It wasn't coming from the outside perimeter so I was confused. I thought it might be coming from a section of the house that had been added on and that the roof probably had a leak in it. I started checking all over the house. Suddenly water started coming in from all over. Outside, the water was getting to be about three inches deep and I worried about my new car. Right in the middle of the storm, I got in my car and drove it to a higher section of the property into a shed that appeared to have no water in it. I thought as I went back to the house that I would be blown away by a strong gust of wind, but again, the wind was not that bad nearer to the ground. The water quickly began to get deeper inside the house. Since I was in the process of moving in, I had many boxes packed with stuff in the garage. Hurriedly, I decided which boxes had the most important stuff and brought them upstairs. Mostly I was concerned about all of our art work, writings, photograph albums and Mama's boxes of genealogy papers. My parents had spent many years gathering information about our ancestors, and had traveled to France and Spain to do research before my father died in 1980. I knew that these were important and didn't want anything to happen to them. I put as many of the boxes I could higher up on chairs and tables and thought I had most of it moved out of the way. But the water kept rising. Cherie helped me as best as she could to move some of the furniture, but it was no use, the furniture was too heavy. When the water got to about ten inches deep, it finally stopped. It couldn't have come from the rain because it came too fast, so it must have come from the lake. My property was only a couple of miles from Lake Pontchartrain on the north side. In fact, the St. Tammany Parish area is called the North Shore. I believed that I was far enough away from the lake that I would not get any water. But I guess I was wrong, at least for this hurricane. (I was to find out later that the water did not come from the lake, but from the river nearby that overflowed.) There seemed to be nothing left to do so we went upstairs to the second floor and continued to watch and wait for the storm to recede.

I reminisced about the fact that I had just moved in and was getting settled, that I had just gotten the first floor fixed up the way I liked it, hanging pictures and putting furniture in their permanent places, setting up the kitchen the way I liked. When I bought this house, I decided that I was never going to move again. I felt that I belonged here.

Rita and I loved to decorate. I was older than she was, and we grew up in different eras; me, the Elvis years, she, the Beatles. Although we were practically opposites in personality, me being the quiet slow one, and she the outgoing hurry-up type, we both had the same interest in writing and art. We had fun fixing up a large game room that was previously a garage, into four sections; a library, a music area, an antique area and the laundry area. We planned a going-away party for Lauren and Kaylee and had French things and colors all over. We put up our art work and photographs all over the walls. The house was spacious enough to have a big party without people being crowded in small rooms, so we invited lots of family and friends. That was two weeks before the storm. Family and friends came and were enjoying my new house remarking on how lucky I was to find such a wonderful place. It was a time for friendship and joy. My new home was looking great and I was feeling proud. And now what? Was my beautiful space, my home, my sanctuary, my pine trees, my pond, my peaceful environment going to be destroyed before I even got a chance to settle in? Was my dream of a quiet place to paint and write in my retirement years going to end before it even began?

2.
Candles in the Darkness

By Monday afternoon the winds were letting up, the rain stopped and the water level appeared to be dropping. I found an old broom and started sweeping the water out of the house. The newer section of the house was a few inches higher than the rest, so I started there. The water was above my ankles and walking through it was not easy because the floor was slippery and muddy. I didn't know how long I was sweeping, but my shoulders began to ache. I kept sweeping and sweeping until I could do it no longer. I would stop often to drink some bottled water. I wasn't hungry, but I knew I that should eat something, so I ate some of the snack food, which wasn't very nutritious, but was all that we had. Then I would go at it again. When it seemed that the water had gone down enough on the property, I started to walk down the drive to the highway. The distance from the highway to the house was about a block long. My driveway was a gravel road off the highway, which splits into two sections, with a cross-over section in the middle. The road was surrounded by majestic pines trees, magnolia trees, bamboo reaching up to fifteen feet or more, and other trees that I couldn't remember the names of. Many of those trees, large and small pines, lots of branches and bamboo were blocking both sections to the exit. There was so much debris dispersed among the downed trees that the walk to the highway was like trekking through an uninhabited forest. This must have been what the pioneers felt like as they made their way through the forest to inhabit this land. I climbed over branches and under fallen trees, finally making my way to the highway. Highway 190 didn't look any better. Trees and more debris were criss-crossed over the road as far as I could see. Someone must have cut their way through the trees on the highway with a chain saw making an obstacle course large enough for a car to drive this way and that to get through. Then I remembered hearing the sound of a chain saw earlier, but did not pay much attention to it. Realizing the intensity of the damage, I began to worry about Cherie needing to get to dialysis and I wondered how I would be able to get out through the mess to get her to a hospital or dialysis center somewhere. I knew that there was a fire station about a mile down the road and thought that if worse came to worse, I would go there and get help. I still had two or three more days to solve that problem. I even planned that since my niece Alyssa's truck was parked here, I would try to tie a rope to it and the trees to pull them out of the way.

As I wandered down the highway, there were a few cars and trucks making their way around the cut trees. I decided to turn back and saw Cherie by the gate looking agitated. She had followed me to the road, but fell into a ditch. A young man was also on the road offering cold water to the passersby. He was said that he had cold water, ice, a bathroom and a shower because he had a generator. We accepted the water and discovered that he was the one who cut trees down to make his way home after the storm. I told him of our dilemma and he offered to cut the trees out of my driveway. Cherie and I spent the rest of the day sweeping out the water, and clearing the branches from the driveway. The heat, humidity and mosquitoes slowed our progress, and by nightfall, we could do no more. But as tired as we were, and with the night uncomfortably steamy, sleep would only come in short bouts. All day Tuesday, Cherie raked the driveway and I swept more water out of the house. We would take breaks at Hank's house for cold water and contact with another human being. We were able to plug in our cell phones and tried to call Nicolle and Rita. When the calls were not going through, we tried voice mail and text messaging. Finally something worked. "wher R U" got an answer: "we OK." "when U comin bak?" "Cant no gas." We didn't want to risk sending long messages, not knowing if we, or they, could re-charge our phones, so we settled for short messages. At least we knew each other had made it through the worst. Or was the worst yet to come?

About noon, I decided to take a ride down the highway to see if we could get to Rita's house ten miles away in Mandeville. I had to drive from one side of the highway to the other and sometimes on the shoulder, just to avoid the fallen trees and debris. Other people had cut part of the trees in order to be able to drive along the highway, which made the drive like an obstacle course. I got as far as Fountainbleu State Park, which was about half way to Rita's house before we could not go any further. No one had cut the trees from that area, so we had to turn back. There were a few other cars on the road, everyone nodded solemnly at each other as we passed each other by. The two small rivers we crossed had water up to the bridges. There was still so much water on the sides of the road that we could not tell the river from the land. There were no stores or gas stations open, and hardly any people except for the few cars. We could hear the sounds of chain saws and generators. Other than that, there was nothing. No phones, no electricity, nothing. We were isolated from the rest of the world.

The nights were still, dark, and hot. We only had the radio that we listened to occasionally, but all that they seemed interested in was reporting what was happening in the Superdome. Then, it was on the radio that we heard that St. Bernard Parish was completely under water. Nicolle's house was in St. Bernard Parish. I wondered how true those reports were and worried about Nicolle and her family. Had they heard it too? Did they know more since they had a TV in their motel. Were they still in Florida, or had they tried to come back? It was on the radio that we heard that Lakeview was flooded because the 17th Street canal levee broke. That was the canal that separated Orleans Parish from Jefferson Parish. My aunt and uncle lived a few blocks from the 17th Street canal on the Orleans Parish side, and I worried if they had evacuated. I hoped that my cousins had gotten them out, but there was no way of knowing. It was on the radio that we heard that Metairie had flooded. My old house was in Metairie and I had not sold it yet. I wondered if it had flooded. My other aunt lived alone in Metairie, not far from my house, and I worried if she had evacuated. It was on the radio that we heard that all of New Orleans had flooded. It seemed that these reports could not be true. All of these areas of Greater New Orleans were reportedly under water? I could not imagine that this was real. The news media must have been exaggerating. There was probably only street flooding, I thought. There is no way that the entire area was inundated up to the roof tops. Cherie and I began thinking of the people that we knew, family, friends and those that I worked with, wondering if anyone was safe or had stayed. Cherie and I would occasionally say to each other, "What about so and so? Do you think they are ok?" Most of the people that we thought of had not evacuated before and some of them had said that they would never evacuate. There was no way to find out about any of them. All I knew was that my immediate family and I had made it out of the city alive and for that I was thankful. But I still had no idea where they were and if they were still safe. From what we heard on the radio of total destruction and death, I wondered if I had any other family or friends who made it out alive. Was this some horrible nightmare that I was having? Would I wake up soon relieved that this didn't really happen? It was too astonishing to believe that the reports we were hearing were real.

My immediate responsibility was to see if we could get out of here and drive to a hospital. I felt confident that I would be able to do so and kept telling Cherie not to worry. I hoped that I appeared confident to Cherie. She was not as confident as I was. She had been through so many disappointments in her young life, since she developed Lupus at fifteen years old, that she couldn't help believing this would turn out bad also.

Wednesday morning we planned to go to Slidell and find the hospital. The problem was getting out of the driveway. The part of the driveway that I used as a crossover was very low and still had muddy water that I might have gotten stuck in. I had already used Alyssa's truck to go out on Tuesday to try to go to Mandeville, and it made it through the mud, but the truck was not very reliable because it was old. My other niece Lauren's car was there also since she had gone to France, but it was also old and unreliable. Cherie's van was not an option because she had so much stuff in it that there was no room for me. She also had OCD (obsessive-compulsive disorder), so she saved things, lots of things. It would take her hours, even days to clean out part of her van. That left my car, which had a half tank of gas in it, and my mother's car, an old Lincoln. I chose the Lincoln because it was a strong, heavy, and partly reliable car, and the Slidell hospital was less then ten miles away. I saved my car for a "real" emergency. The Lincoln kept getting stuck in the muddy part of the cross-over even though I had put some old rugs and welcome mats in it to improve traction. Fear returned as I tried to go forward, then backward, again and again, and still couldn't get through. I began wondering if I was doing it right, until finally I backed up far enough, drove through at top speed, tires spinning through the mud, and crossed the muck to the other side. I managed to drive to Slidell down Highway 190 to where I thought the hospital was. I was only guessing since I was new to the area. The drive was like being in a sci-fi movie. There were downed trees, branches, twisted metal pieces of roofs and buildings, fallen buildings, boards, downed power lines across the road and along the sides of the road, stuff everywhere. But not many people. Someone had cleared most of the debris off the road, probably the Army or National Guard, so it was passable. I had heard that help was being sent here. The hospital was where I thought it was and I believed that this would be the end of my worries. I asked someone directing traffic, (what little there was), where to go and he pointed in the direction of the emergency entrance. There were a few medical people standing outside so I drove up to them and told them what we needed. To our disappointment they said that they could not help because they were not equipped to dialyze anyone. They had no electricity, not even generator power or clean water, which was essential for running a dialysis machine. But the nurse directed me to a triage area down the block. The triage area was nothing but a parking lot with covered parking spaces next to what looked like an apartment building. People were lying in the shade on the ground on thin blankets, and little children were crying. They were treating only serious injuries. I was reminded of the MASH units from the TV program and expected to see something similar. But *MASH* was only a TV program, and could not have captured the reality of a true wartime situation, or of a devastating event like this. This place was nothing like *MASH*. No tents, no doctors treating people, no operating tables, just an open area with injured people lying on old blankets. The nurse there said that we would probably have to drive to Jackson,

Mississippi to find an open, clean unaffected hospital or clinic. They were unable to give us any real information about where the nearest clinic was that was open, because there were no phones or cell phones that worked, not even emergency lines. Communications were down everywhere. I asked Cherie if she wanted to drive to Jackson from there and she said no, that we could go tomorrow since she wasn't feeling too bad. She had hardly eaten anything the past few days and the water she drank was eliminated through sweating. We returned home to do more cleaning and plan for the next day's trip. I parked the car on the other side of the crossover so that I wouldn't have to worry about getting stuck.

I wanted to get an early start Thursday because I thought we would have a long way to go, but Cherie took her time getting ready, as she always does. She was determined to rake a few more leaves from the driveway before she packed for what we thought would be a long stay. I learned not to "push" her because that would only get her frustrated and then she would take longer to do things. So I told her to let me know when she was ready, and I went about doing more cleaning. Both of us dripped with sweat during the day when we cleaned up and even at night trying to sleep. We slept with the windows open hoping to get some semblance of a breeze, but there was none. The days and nights were quiet, hot and humid. I didn't even have running water to cool off with. When I found out that the property had well water, I was pleased, thinking that I would not have to pay a water bill. But the reality was that the water in the well had to be pumped up, requiring electricity. I knew that I was supposed to save some water before the hurricane, so I tried to fill the tubs with water, but the water leaked out of both of them. I did have two ice chests full of water though, but I would not use that for drinking because it smelled like egg water. At least we had enough bottled water to take with us.

Finally she said that she was ready to go. We packed enough clothes for a few days, what snack food I had, and some bottled water. I locked up the house and left a note on the door in the event that someone might come looking for us. I decided to use my Nissan Altima instead of the Lincoln. We went to Hank's first to tell him where we were going. He asked if we could get him some gas for his generator, and mail a couple of letters to his family to let them know that he and his children were ok. I gave him some of the snack food we had and a couple of children's books I had from teaching. It was a odd sort of good-by, strangers thrown together by this event, but concerned for each other's safety.

I decided to go west instead of north. I had been to Jackson before. It was about a three hour drive with not many towns in between that I thought would be open since the hurricane took a northerly path. I had also been to Houston. There seemed to be more towns on the way there, and I planned to stop at each one of them until I found a town with an open gas station and/or a dialysis clinic, or at the very least, communications. With a half tank of gas, I had to choose the path that gave me the best options. Since many people had evacuated to Houston, I knew there would be no vacancies anywhere along the way, but I had two relatives and a friend there and felt confident that one of them would let us stay with them.

I went up to I-12 and drove west. I went about forty miles before taking the chance that something would be open. I saw a hospital sign on I-12 in Hammond, so I drove off to find it. It was only a mile off the highway and was easy to find. I drove to the emergency room entrance and explained our situation. I was given the same story as in Slidell. My heart sank. I thought we were going to have a long way to go to get help. The feeling lasted only a minute as she told us that there was a dialysis center just across the road. We went there hoping for better news. They were open and were accepting patients, but were full. They were running on a generator and had many evacuees to take care of. They told us that we could go back to Covington or further west. Again, disappointment, when we were so close.

But as we were trying to decide what to do, a nurse came out and said that they were opening an 11:00 (PM) shift. It was only 4:00 in the afternoon. We could stay or we could drive on. We decided to stay since I was low on gas. Besides it would be a sure thing if we stayed but more uncertainty if we left to find another place. We told them we would definitely be back at eleven and they put her name down. We had about seven hours to find something to do, but at least I wouldn't be using more of my precious gas.

Hammond had also gotten damage from the storm and had no electricity anywhere in the city. A couple of gas stations were open using a generator, but nothing else. We drove back to the hospital where we could at least find a bathroom. We started looking for a cold drink or something to eat, but the snack machines ran on electricity and the cafeteria was closed. It was slightly cooler there, but Cherie was getting desperate for her root beer and air conditioning. She wanted to go look for another place that might be open. I was getting annoyed. Didn't she realize that there was nothing open anywhere? She started acting like a child throwing a temper tantrum because she couldn't get what she wanted. I was getting mad. She was getting mad. We had been together for five days. We were hot, hungry and hadn't had a shower, or a good night's sleep. We were worried about our family, where they were and what they were doing. Maybe the poisons were building up in her body and making her irritable. We left the hospital to find a motel so we could get some rest. There was one behind the gas station that actually had a room, but they had no electricity. I thought maybe I could take a cold shower and rest awhile, but they were only taking cash. Their credit card machine needed electricity to run accounts through. Well, I had no cash. I didn't have the foresight to cash a check before the hurricane, or fill my tank up. I was really not prepared for this. We were becoming more annoyed with each other, so I drove Cherie back to the dialysis center and left her there. We needed to get away from each other for a while. At least at the dialysis center there were people that she said would understand her and she could talk to. She didn't think that I understood her problems. Of course I understood. I just couldn't DO anything. I said that I could not make electricity for her to be cool, or find a cold drink just because she wanted one.

After dropping her off, I drove around to be by myself for a little while. I didn't want to go too far, but I wanted to find a motel. I checked out a couple more, but they were all full. A couple of hours later, I went back to pick her up. The hospital said that they would open the cafeteria for employees and visitors at six. Maybe we just needed something good to eat since all we had was bottled water and snack food for five days. I told the people that Cherie was talking to in the dialysis center about the cafeteria and they decided that they would go too.

The menu was sparse, but it felt like dinner at Antoines. There were only three choices: spaghetti from the day before, pork chops, or a sandwich, each with corn, a roll and a cold drink, all for only $3.00. We got two plates of spaghetti since it was a hot meal. She didn't like the spaghetti and went back for the sandwich. She paid for it all because I didn't even have that much cash. I ate all of mine, plus the spaghetti that she didn't want. We looked like we hadn't eaten in days, which was true, and like we hadn't showered in days, which was also true. It would have been embarrassing except everyone else looked the same. Bedraggled.

As we were finishing eating there was a small commotion. (There were hardly enough people there to have a big commotion.) Their air conditioning had just come on. Ahhhhh, cool air! A good sign! We still had more than three hours before dialysis so we hung around the hospital after the cafeteria closed. We sat in a waiting area where there was a place to plug in our cell phones. Occasionally, a security guard would eye us suspiciously and then ask us what we were doing there so late, and I would tell them. One of them explained about National Security and having to be on guard, and would then leave us alone. Did we look like a couple of dangerous people? Two women, tired, hungry and dirty? By 10:30, we went to the dialysis center and they took her right away. A dialysis patient usually runs for three to four hours depending on how much the person weighs. Cherie weighs only 115 and runs for three hours making it 2 AM before she would get out. I thought I would take this chance to take a nap. It was cold in the lobby since they had their electricity back. I couldn't sleep. There were a couple of other people there talking about their experiences. One older lady who was waiting for her husband was from hard hit Slidell, just five miles east of Lacombe. She was telling of how they stayed in their house and felt safe until a tree fell through their roof. The rain came in through the hole and flooded them. They stayed in the bathroom because it was supposed to be the safest room in the house. They spent hours and hours huddled in the bathroom until the storm let up. After that, they had to wait until the water went down and the road was clear before they could go anywhere. They spent a couple of days in a house that was destroyed. When they did manage to get out, there was no place to go. Slidell had been hit hard and had no electricity, no businesses open, no gas stations in the entire area, and no open dialysis center. They couldn't drive to their nearest relative, nor call them. They spent nights in their car. Then the social worker at the dialysis clinic found them a place to stay until they could be re-united with family. I knew that there would be thousands of stories like their's. I felt sorry for the older people like them who had no one to help them. No matter how uncomfortable we were with the heat, at least we didn't have to sleep in a car. Yet.

I began thinking of all people who had no one to help them, and realized how well off we were. The reporters on the radio were telling of the rescue helicopters that were finding people on the roofs of their houses days later. They were rescuing people by boats because so many areas were still flooded that they could not get even the large trucks through. The reports on the radio and the TV in the clinic lobby were so depressing claiming that thousands of people probably lost their lives and that the government was preparing to send 25,000 body bags to the area. This ominous attitude combined with the unbelievability of the state of affairs had me feeling numb. I could see the same melancholy in the faces of everyone we met. We were all in this alone together. It was truly an "every man for himself" situation. The destruction was so massive that individual needs could not be met. The people that were sent here to help were so overwhelmed that it was hard to know what had to be done first. An occurrence of that magnitude had never happened before. Officials led us to believe that if we followed their instructions about evacuating and having water and batteries and food for a few days, that they could handle the rest. We were led to believe in a false sense of security. It wasn't exactly their fault. They could have had planning sessions, meetings, input from all types of agencies and businesses and individuals, but until it really happens, "this is not a drill," there is no way that anyone could have known what problems would occur. We, as individuals, trusted the authorities, believed that if we got into trouble that someone would be there to rescue us. If we followed their advice, everything would be ok. But that was not happening. We were alone. Whether we evacuated or not, we were left with our own resourcefulness to solve our own individual problems. My sister evacuated, but there was no place to go. Thousands evacuated, but there were no places to go. The rest of the country was not equipped to handle half a million evacuees. Those of us who stayed had to rely on our own ingenuity to handle individual problems to save ourselves. I had to get my daughter to a dialysis center but there were trees in the way of my path. I could not call a local tree-cutting service to come over and do the job. If Hank did not have a chain saw, what would I have done? If I would have used Alyssa's truck to pull the trees out of the way and got nowhere or worse, broke the truck, what would I have done? Was I just lucky to have a neighbor with a chain saw, or did I make my own luck because I DID something, I tried. Someone once told me that if I was faced with a difficult decision, to do SOMETHING, do ANYTHING, even if it would turn out to be wrong at least I tried. If I did nothing, would we have become one of casualties like so many others? Did casualties occur because people didn't try? We really were pioneers in the darkness in this new age of technology. No matter how great the country thinks it is with computers and cell phones and instant communication with the world, we were nowhere, cut off from the world, left to our own initiatives. "Help is on the way," means nothing when people are sitting on top roofs for days, or struggling for breath in an attic, or sleeping in cars for days, all with no food or drinkable water. Was it our fault? Should everyone living in a potentially dangerous section of the country move to Arkansas, because nothing ever happens in Arkansas?

At 2:30 AM, Cherie came out to the lobby. It was too late to drive back, besides I needed gas. The stations would not open until seven that morning. As we drove toward I-12, we noticed that the Waffle House right by the interstate had lights on. We went in to find that they were indeed open and running on a generator but had a limited menu. That suited us just fine. We could actually have waffles and coffee! Coffee! What a treat! I realized that I hadn't had coffee in five days! So that's why I had a headache. I ordered a waffle and Cherie ordered the eggs, toast, and grits special. We planned to split our food so we could both have some of everything. I mentioned that I would love a pecan waffle but Cherie did not like pecans. When the waitress came with our order, she had an extra pecan waffle explaining that they had "accidentally" made an extra one. Did we look that pathetic? We ate everything but half the pecan waffle. We sat there for over an hour because there was nothing else to do and nowhere to do it.

We listened to the conversations of the police in the next booth discussing the goings on at the Superdome. Was that the ONLY place anything of interest was happening? How about the other half million people dispersed throughout the South? I put the half of the pecan waffle in a to go box, (I was not about to waste any food) and we went to the car to try to sleep. The Waffle House was right next to a Pilot gas station where many truckers stopped for gas and food. This one was closed for the night because they only had generator power and not enough staff. That was the situation everywhere; no power, no people. There were lots of cars parked in the bays with people sleeping, waiting for the gas station to open, or just because there were no motel rooms to sleep in and it was a safe place to be. It was 4:30 AM and I was just about to doze off when the flashing of police car lights woke me up. I watched as the police got out of his car, and I wondered what was going on. I guess he was just checking out the area, but he left his car door open and someone's dog jumped in the car. I had seen this man walking his dog around the area before we ate, so he must have been one of the "sleepless in the gas station" people. The dog's owner and the policeman tried to get the dog out of the car, but he wouldn't move. I thought that the policeman would be angry with the owner, but what could he do? Arrest him because his dog made an illegal entry into a police car? Why did I think that the situation was funny when I was sleeping in my car in a gas station at 5 AM, my house might have to be gutted out, and I had no idea where my family was nor did I know the fate of the rest of my family and friends? After the dog was out of the car, I couldn't get back to sleep, so I watched the people. There was a hefty black woman approaching the men as they went into the Waffle House. She would talk to them and bum cigarettes from them. Then she sat on the bottom of the light pole until she needed another cigarette. One man offered to buy her breakfast and she went in to eat. She brought out half of her breakfast in a to go box and started walking away from the gas station. Later she returned wearing something different and began her routine again. Was this a regular thing or was she another victim of the hurricane?

At 8:00 AM the gas station opened. I had to go around to the long line that had already formed in the roadway. I knew we would be there for at least an hour. I hoped they would accept credit cards because I didn't have any money. Cherie would be able to cover it if I couldn't charge it, and I would pay her back later. They were only allowing people to fill up one extra gas container due to the shortage, so I filled the biggest one for Hank. After that, Cherie wanted to drive to Walgreens nearby. They were open, but their parking lot was full of big orange plastic pieces from the McDonald's sign across the highway. Since they were running on a generator, supplies were limited. There were no cold drinks or anything that could spoil. We settled for more snack food and started for home.

3.
Reunion

It was almost noon by the time we were on the way back home. I wanted to get off the Mandeville exit to check on Rita's house and her cats and dog. This road was also an obstacle course, with trees and power lines and other debris scattered all around. Going around the bend towards her house I could see huge pine trees that had fallen on house after house, some split right through the middle of them making a clean precise cut, some just through the sides, and some houses completely undamaged. What was the reason for this random act of Nature to chose one house to demolish and not another? Or one life to end and not another? I feared the worse for my sister's house. As I turned the corner of her street, I thought I saw her car pulling into the driveway. Yes, it was them! Both of us arrived at the same time. We hugged and cried. We stood there and stared. The unbelievability of it all had frozen us in the moment as time stood still. There seemed to be no words to express what we felt. We were five women, Rita and I, Mama, Alyssa, and Cherie with her chronic disease, and we made it through the worst storm in recorded history, and here we were, reunited, and safe.

Her house was one of the lucky ones with no significant damage that we could see, and her animals were ok. There was no electricity, though. We caught up on our adventures of the past few days and laughed at the stupidity of it all. They had gone to my house first and read my note, but did not know where we had ended up. They left a note of their own so that when I returned, I would know that they were back home.

"You won't believe what happened to us!" Rita said.

"Did you stay at Linda's in Alexandria?" I asked.

"Yes and no." she answered. "After we last talked on the cell, we drove to Linda's. It was after midnight and we were afraid to wake them up, but we did. They had people sleeping all over, in all the beds and on the floors. They had two cots ready for us and Mama went right to sleep. But when we brought the cat in, Linda's sister said that we could not have the cat in there with all the people and dogs, that the cat would have to stay in the car. Alyssa and I looked at each other and knew what each was thinking. We were livid! We couldn't stay there."

"So what did you do?"

"I woke up Mama and told her we have to go. But Mama said 'NO! Just a few more minutes, please!'"

"You mean you left?" I asked incredibly.

"When I tried to tiptoe to the bathroom, the two dogs that were there woke up, barked and then there was a big commotion. Everybody was awake and we felt more uncomfortable. Linda and her sister begged us to stay, even said we could let the cat in, but I told her we would go to Baton Rouge to find Gary (Rita's ex)." "I lied, knowing we wouldn't find him, but we couldn't stay there. So we got back in the car and drove away."

"So where did you end up?"

"We drove back to Baton Rouge. Mama was snoring, and we all had to go to the bathroom. I almost fell asleep because the lampposts reflecting on the road were hypnotizing. But I was determined to find a motel! No one had any vacancies. Then I saw the reliable Marriott where I had stayed before."

"What time was that?" I asked.

"It was about 5 AM, and I begged them for a room. I told them a cot in a closet would do. We had driven all night and I had my 86 year-old mother in the car. Do you know that they actually came out to the car to make sure I really had an 86 year-old person with me? Then the desk lady said to wait while she went to the back to talk to the manager. When she came out she had a room! Alyssa was laughing when I got back to the car."

"What was so funny?"

Alyssa said, "While we were waiting in the car, a man from a TV station shoved a camera and a microphone in my face and asked me about our ordeal."

"Why was that so funny?"

"Because my hair was a mess!"

"So we just left this morning and went to your house, but you weren't there." Rita said.

"I left you a note."

"We left you one too."

The whole situation sounded like a Chevy Chase vacation movie.

After I told them what happened to us, we asked each other what we knew about other family members that lived in the flooded areas. We hadn't heard form anyone, not even Nicolle. All we knew was that St. Bernard Parish was entirely under water, along with New Orleans, New Orleans East, Lakeview, and parts of Metairie. We had family and friends in all of those areas and didn't have a clue what happened to them.

We ate whatever was there that was still edible, mostly bread and cheese. Then I took a shower! The first one in five days. Even though I don't like cold showers, (no electricity meant no hot water!) this one felt wonderful.

Although there was no electricity, Alyssa noticed that there was a light on the phone. "There's a stream of electricity, see that!" she said. About an hour later, the phone rang. What a foreign sound in a house of silence and darkness. It was our aunt who lost her house near the 17th Street canal. She was able to tell us about our family members that had evacuated with them. More family alive and accounted for. With that good news, we were able to sleep a little better that night, even though with the windows open, the air inside was as hot and humid as the air outside.

Friday morning I drove back home. First I stopped at Hank's to deliver the gas. He didn't expect us back so soon and was so thankful because he was nearly out of gas to run his generator. Mama, Rita and Alyssa came over to see if they could do anything to help clean up. I had no idea where to begin. The water was out of the house, but the floors were slippery and damp from the slush that remained. While looking over the damage, they discovered that I had not saved some important papers that were in the bottom drawer of a file cabinet. The water had gotten into it, and into some boxes that I thought were not important. All of the downstairs furniture was damaged because of the water, including two antique cedar chests that had collectibles and a couple of photo albums in them. We opened the albums and peeled the pictures out, then laid them and the papers out on the gravel road to dry. I tried to mop the kitchen by spraying cleaner on the floor and trying to mop it up, but had so little water left in the ice chest that I only got a small section done.

As we stopped for a break to eat something, Mama noticed a candle on the table outside. For years, she had been buying these tall, religious candles to keep lit on the stove. They would last about a week, then she would throw it away and light another one. When she moved in with me, she kept up the same practice. These candles had religious figures on the front of them, are inexpensive and can be bought at K-Mart or Walmart. One morning, about two weeks before the hurricane, I was getting my morning coffee and noticed that the candle was burning black smoke. Not thinking anything of it, I mentioned it to Mama. She got all upset and said, "Blow it out immediately!" Well, ok, I did that and asked what was wrong. She said that that was a bad omen and I had to get rid of it. In all the years she had been burning candles, none had ever burned black smoke before. After I blew it out she lit another one, but I did not get rid of it. I kept it. Everyone we told that story to agreed that it was truly an omen that something bad was going to happen.

Since we were discussing omens, I remembered a dream I had about the same time, two weeks before the hurricane. My son was in the dream. On March 15, 2005, Ricky had taken an overdose of Somas and other drugs. He had been having problems with drugs and depression, and we all tried to help him. He was hospitalized and was seeing psychiatrists, but nothing seemed to do any good. He was put on prescribed medications, but they would only help for a few weeks, then he would go back into deep depression again. He would be given stronger and stronger medications but the same thing would happen repeatedly. I thought that if he would move with me to my new house, he could recover and start over in a new environment. He seemed to like the idea, even saw the house with me and planned which room would be his. He would be able to help me take care of the acreage and I could help him get a new outlook on life. We all thought he was anxious to try this new plan. But we were wrong. We saw the signs, but it did no good because the help that we were supposed to get from people in charge never materialized. The only answers that we got from the professionals were that if Ricky wanted help he would have to put himself in extended treatment, that we could not do it for him. But he would not go. He said he "could quit anytime he wanted." He chose to quit life rather than quit drugs. They called it suicide. I said that he overdosed and maybe didn't mean for it to be fatal. We will never know which answer is right.

Since then, I have had many dreams of him. In the dream two week before the hurricane, I dreamt that I was leaving a medical building and four men dressed in white hazard suits told me to go west. I went down a hill to the other medical building that I was previously in and wanted to go inside to get my raincoat. But the building was evacuated and people were standing outside wondering what to do. Ricky called me on my cell phone and told me not to return to my Metairie house because it would blow up, but to go west. I asked him if he could see the future because I had a racing form in my hands and wanted to know who would win the next race. He answered, "Kory." (Kory is the name of a friend of his. When I told this dream to Rita, she reminded me that Kory begins with a K like Katrina.) After hanging up the phone, a young Nicolle was with me and I needed to get a younger Cherie and Ricky out of the day care that they were in. Just then, a school bus drove up and Cherie and Ricky got out of the bus. I was glad that we were reunited and could go west where we were told to go. That was the end of the dream. I had been discussing dreams and their images and meanings with a therapist. This one seemed important at the time and was the topic of one of my later sessions. Whether anyone believed in the supernatural or not, that dream had some powerful imagery that could be associated with a hurricane called

Katrina. It could even be interpreted as prophetic. What, I wondered, is the point of having a prophetic dream if I could not use the information to anyone's advantage? If I could or did try, no one would believe it anyway. If I did understand the meaning of the dream, would I have done anything different? I was thinking of the dream when I chose to go west toward Houston instead of north to find a dialysis unit for Cherie. That reason, and one as simple as a butterfly is what led to my decision to drive towards Houston.

After Ricky's death, I went to a group called Compassionate Friends, for people who have lost a child for any reason. Their symbol is a butterfly. In May of 2005, they had a butterfly release in Lafreniere Park, near my Metairie home. I took some black and white photos of the butterflies and when I got them developed, one of the pictures had the image of a cross next to the butterfly. What could that mean? I didn't remember seeing that when I took the picture. The image is that of the butterfly on someone's finger, and the cross is in the background. It was a very hot day and someone had watered the plants nearby. Water was still in the cracks between the bricks in the shape of a cross. That showed up on the picture. I couldn't understand why the bricks were still wet since it was so hot. The butterflies that flew around me in Lacombe reminded me of the butterflies on that day in Lafreniere Park. As I was deciding whether to go north or west, one of the butterflies flew nearby and landed on a chair where I was standing. It stayed there for quite a while, then it took off in a westerly direction. Was I being given a message? I made my decision on a belief that it was a message. Were all of these things just coincidences? Is the subconscious somehow linked to the supernatural, sending us messages to help us prepare for traumatic events? Because people don't have any answers to this, they either laugh or scorn the idea. Behind their laughs and scornful looks, many believe but do not want anyone to know that they do. In my recent readings to find answers that will help me to make sense of this life, I read that there are no such things as coincidences. There are no accidents. Everything happens for a reason. We are where we are supposed to be at all times and come in contact with people we are supposed to be with at all times. In my struggle to understand this concept, was I interpreting every little thing as a sign of something bigger? Why things happen to people is a question even the philosophers and theologians can't answer, but I need to know. My daughter, Cherie, has an incurable disease; my son, Ricky, took his own life; and my daughter, Nicolle, lost her home and possessions in this devastating hurricane. Why?

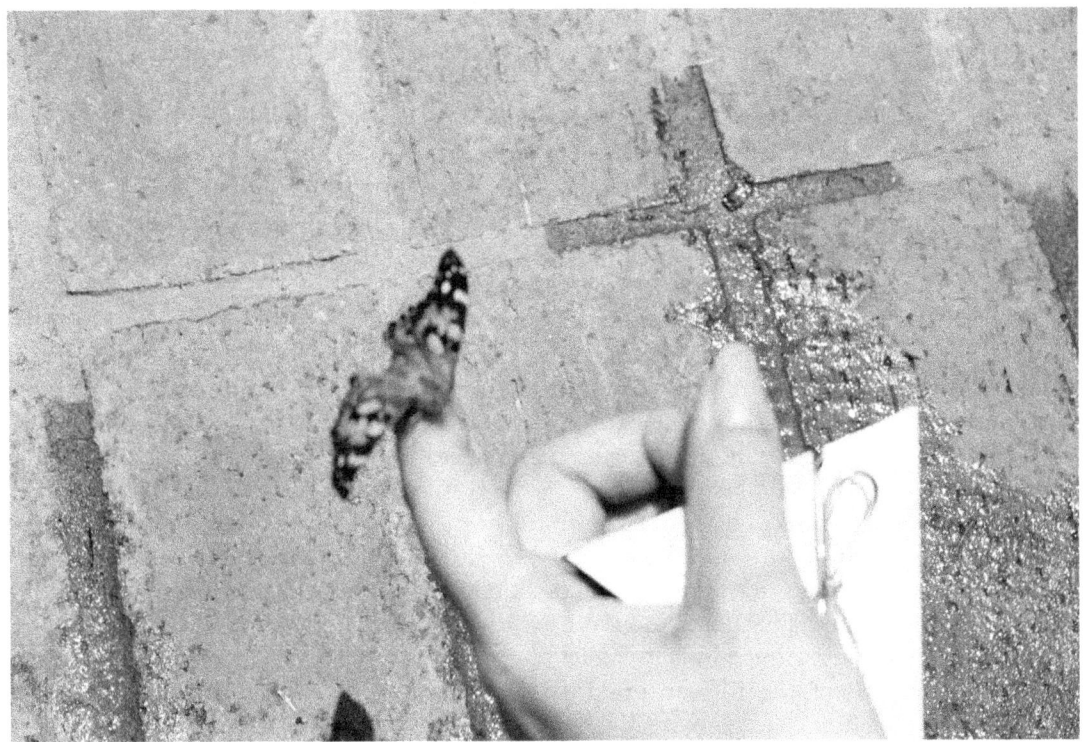

Butterfly and cross

Friday night, I drove Cherie back to Hammond for dialysis. They put her on a schedule of Monday, Wednesday and Friday at 11:00 PM. While she was there, I drove to the nearest motel that had lights on to see if I could get some ice. There was nowhere to buy any, so I decided to get some from the motel, even though I was not a registered guest. At that time of night no one would notice me. Being paranoid, I felt guilty. The reports of looters in New Orleans came to mind. People were stealing food because there was nothing to eat and nowhere to get anything. In desperate times people did desperate things. Five days after the hurricane, resources were still scarce. There was still no electricity, no communications, and no help. Ice was a major necessity, not a minor need. The machine at the motel was full. I knew that if I asked, they would turn me down saying that the ice was for their paying customers. So I justified my decision to take some ice because I was needy and they had plenty.

After picking up Cherie, we noticed lights on in a small business along the highway. It was a bar. Imagine that! No one in the entire city can open for business, but at 2 AM, there was a bar open with a number of cars in the parking lot. Well, I guess, priorities rule. The most important thing for some people at that time was to get a drink. Why not? There was not much else to do. It was that or the Waffle House. Or both. We were thankful that there was a place to get some "normal" food, especially at 2:00 in the morning. We went there again and brought coffee and waffles back to Mama, Rita, and Alyssa. The simple pleasures of having coffee and ice were all that we needed to be comfortable and satisfied for a little while.

Our cell phones were still not working right. I had been charging our phones at the dialysis center since their electricity was back on, but we were afraid to use them too much because we didn't know when we would be able to charge them again. We were still text messaging. Cherie's boyfriend messaged that Cherie's trailer was ok, no damage. That was one less thing to worry about. On Friday, I had not heard from Nicolle in a while so I text her this message, "We r ok come home." And again later when I still hadn't heard from her, "Hey wher r u? We ok cant call." On Saturday, Nicolle was able to send me this message, "I notcd. We dcidd 2 leav in morn his frnd has xtra house we cn stay in 4 as long as we ned. Kids wil b abl 2 go skl rit way cl me wn pho cm on." Later, when she got to Baton Rouge, she sent this message, "Yes I ben trn 2 cl U. Cc cl me tis mrn did I gv u # hre in btn rg." Even this small contact with Nicolle was reassuring. I knew that she was in Baton Rouge staying in a house of a friend instead of in Pensacola in a motel.

Cherie had decided to drive to her trailer since it had no damage. She lived near LaPlace just at the end of I-55. The two shortest routes were open only to emergency vehicles, and that meant she would have to go all the way to Baton Rouge. That would have taken hours. She drove I-55 anyway and they let her through because she told them that she was a dialysis patient and needed to get home. She used that excuse often to get her way. It usually worked.

That same day, Saturday, Rita, Alyssa and I drove to Baton Rouge to find a Laundromat and a bank to possibly cash some checks. My regular bank was in Metairie, but Metairie was still closed to evacuees until the authorities deemed it safe to return. Many parts of Metairie were still under water. The Baton Rouge banks were accepting checks from the disaster area but were limiting withdrawals to $200.00. That was the first money I had in a week.

At the Laundromat, the TV was tuned to CNN with reports about what was happening here. That was the first I had seen of TV in almost a week. They were showing the levee break with the water pouring out of the 17th Street canal into the residential area. My aunt and uncle's house was right there, a block from that canal. The water was over the roofs already and still flowing out. I had heard about it on the radio, but seeing it for the first time made it more real. We thought about all the stuff that Aunt Audrey and Uncle Marion had in that house. All their collections and treasures were lost. All their photos and home movies of all the family, Aunt Audrey's collection of cups, Uncle Marion's memorabilia from the time he was in the Navy until August 2005, their children's things, stuff from all the vacations they took, things I can't even imagine they had, all lost. The reality of what happened to my aunt and uncle, Nicolle's family, and hundreds of thousands of other people over night was an impossible concept to believe.

Then, the reporters switched to what was happening at the Superdome. They were telling of awful things that later turned out not to be true. They were focusing on how slow FEMA was in bringing aid to us. They were already gearing up for the blame game. We didn't want to hear about who was to blame, we just wanted help. Most of the other people in the Laundromat were talking of their ordeals to whoever would listen and were ignoring the TV. We just wanted to plug in our phones and get our clothes cleaned. But the images on the TV were captivating since we had not seen any of this yet. There were aerial views of the invasion of water all over New Orleans and the surrounding area showing hundreds of rooftops in the middle of a sea of water. People were being air-lifted into helicopters off their roofs. A feeling that this could not be real crept into my soul. What was happening to us? Thousands of homes flooded and thousands of people may be dead. There had been so many catastrophic events in the last few years, could this be the predicted apocalypse beginning to happen? Were we all so sinful that we would be totally destroyed soon? Is there a God so disappointed with us that He would inflict this kind of devastation on everyone? I could not then and will not ever believe that a God could be so cruel. If He/She is so angry with us, why make good people suffer with the bad? I wanted to find some logic in life and that was not logical. Perhaps there is no answer, and never will be. Is this our destiny, to always question and never know?

Since we were in Baton Rouge, I tried to call Nicolle on her cell, but she didn't answer. I didn't know the name of the people she was staying with or an address or phone number. She didn't even answer a text message, so that was a disappointment.

On the way back to Mandeville, we went Highway 22 to avoid Interstate traffic, and to get a look at how bad that area was. There were many trees down that were cleared and piled up alongside the road, and many places were not open, probably because their electricity was off. We knew we needed gas. Alyssa was driving Rita's car and she didn't have a chance to fill up after she came home. We passed a small station that was open and didn't have a line a mile long like stations in busier areas, but the line was around the block on a small two lane road. There were ditches on both sides of the road as in most of the areas out of the larger cities. Alyssa turned on the road to get in line, but there was no way to turn the car in the direction of the line. In order not to block traffic coming in the other direction, Alyssa drove backwards into the gas line. No one seemed amused but us. I think we have a unique ability to laugh at situations that other people take too seriously. There, and everywhere else we went there were police directing people into and out of the gas stations and stores. That was a good move on the part of law enforcement, as I am sure that their presence avoided lots of fights and people trying to cut in line ahead of other people. Just like being back in elementary school. No cuts. We got some ice and cold drinks from inside, then went back to Rita's house.

Here we were, five women getting by as best we could with no electricity, therefore no air conditioning, no TV, no internet, no fans, no washer-dryer, no refrigerator, no stove, and no lights at night. So we drove eighty miles to wash our clothes, get hot cooked food, buy ice to keep milk in, and wait in long lines for gas. We slept in the heat, did without coffee, phones or news. We didn't complain because we were alive. We waited it out, day by day, for whatever would happen next. What that was and when it would happen, no one knew.

Luckily, that evening, relief arrived in the appearance of Rita's ex-husband, Gary, father of Alyssa and Lauren. Coming in like Santa Clause, he brought presents of a generator, fans, a huge ice chest filled with ice, water, an outdoor grill, and a tiny TV. Now we were cookin'. He tried to hook up the big TV to the generator, but that didn't work well. (Leave it to a man to put the TV first on the priority list.) He connected a light and the fans so we all found a comfortable place to cool off, and fell asleep. The next morning, Rita hooked up the coffee pot and we had real hot fresh coffee! Such a simple pleasure to start the day with. Then she tried to warm something in the microwave, but the microwave didn't work on generator power, and it blew out.

We went to Lacombe to clean up more. I brought more water in any large containers I could find, like gallon water jugs so that I could continue mopping the floor. We moved the mattresses out because they were wet, then tried to open the bottom drawers of Mama's furniture. The wood was beginning to swell so I had to break the draws open to get inside. Some of the clothes were wet, but not dirty or muddy. We started to move everything that was not damaged to the second floor. The walls were beginning to mold and we didn't want that to spread to anything else. I sprayed bleach on all the walls and everything that we couldn't move to keep the mold from getting worse. Maybe I wouldn't have to throw much away, and could save the walls from being torn down if I cleaned it fast enough.

Labor Day. One week after Katrina. Still no electricity. Still hot and humid. Still no phones or good cell phone service. Still no open stores. Still long lines at the gas stations. It took me four hours to fill up my car and Alyssa's truck. Some businesses, mostly grocery stores, were beginning to open for a few hours a day, but they were short in supplies. Since they were using a generator, there were no perishables. There were police and Army soldiers everywhere. Somehow Gary managed to find some meat to cook on the grill, and ice to refill the ice chest. We would be eating well and had cold water. We cleaned up more in Lacombe, dragging out heavy area rugs that were still soaked and heavy. The odor of mold was getting stronger even though I kept the windows open all day and night. The humidity wasn't helping matters because mold loved dampness. Later, we went back to Rita's house to dine on a home cooked meal.

The next day Rita, Alyssa and I went to Baton Rouge again to wash clothes and check our E-mail on the library's computer. Finally, a connection with the outside word could be established. We were able to get one hour each on the computers. I don't think many people knew to go there yet or it would have been packed. Some people were applying for FEMA help on-line. They were saying on the radio that people could apply for FEMA help on-line, or by phone. Who were they kidding? The majority of us had no phone service or electricity to use a computer! Nicolle was able to get a call through to me saying that she had stood in line all day in the heat for FEMA assistance, and all day the day before to help her in-laws. I wasn't ready to do that. It was too hot to spend the day in line, not knowing if I would be able to get anything. There were so many conflicting reports on the news that no one knew who or what to believe. I decided to wait until things settled down a bit. There were Red Cross lines and food stamp disaster assistance lines. Lines, lines and more lines. Such organization. I would have visited Nicolle that day, but she was in some kind of line for something, and I didn't know where she was. Neither did she.

The authorities were finally letting people go into Metairie to check on their property, but not to stay there. There was no electricity or clean water and it was dangerous to try to move back in. I wanted to check on my Metairie house and planned to go on Wednesday. I knew it would be an all day occasion because there were limited entrances into the city. I borrowed Alyssa's truck again in case there was something I wanted to take back to Lacombe. I got started early and took plenty of bottled water. I went along Highway 51 which follows I-55 to LaPlace, then Airline Highway into Metairie. The traffic was bumper to bumper all the way in. The usual 45 minute trip took four and a half hours. Then it hit me. I left my house keys in my car! Dumb, dumb, dumb! All the plans I made and I forgot the stupid key. I could still find things in the yard to take back; anyway I would get to see if there was any damage to the house.

Metairie, on the outskirts of New Orleans, was a thriving community with a few tall buildings and lots of traffic; close enough to New Orleans to work there, but far enough away to avoid the inner city problems. Orleans Parish, Jefferson Parish, St. Bernard Parish and St. Tammany Parish were each in competition for people, business, and money (as in tax base!) Today Metairie, in Jefferson Parish, was a wasteland, as were the other parishes. Trees were cleared off the main streets, but not the side streets where the neighborhoods were. The wind was so fierce in the hurricane that entire trees, roots and all, were knocked over. Some of the root bases were as huge as the houses that they were supposed to protect. Some pulled up the sidewalk, road, or anything that the tree was planted next to. There were so many downed trees that it looked as if some giant had weeded his garden, removing the foliage from the little box houses. The tree in my back yard was still there, but had lost half its branches. Nothing damaged my roof. The fences were down everywhere. I could see right down the block from my back yard, where before, there were fences surrounding each property. The new shed I had just put up was broken and scattered, but everything was still there. I looked inside the window and it was just as I had left it. No apparent damage. A neighbor said that the water was high, but not enough to go inside the house. That happened before when we had a lot of rain in a short period. The pumps were not strong enough to pump out the water faster then a few inches of rain per hour. We often had flooded streets from thunderstorms. A few houses on my block were lower and got water in them, but this was also usual as they had flooded before. Relieved that my house was dry, I put a few useful things from the yard into the pick-up truck and drove to my aunt's house nearby. She lived closer to the lake on the north side of I-10, and had water in her house about knee high. I couldn't get in, but the water line told the story. All the neighborhoods in that area had significantly deeper water, mostly three to five feet inside the first floor. The water drained out slower than in my Lacombe house, so it caused more damage. Most everyone had evacuated and could not get back to clean up right away, making the damage more severe. Every house would have to be gutted out to be livable again. I didn't know the exact numbers but it would have to have been at least 50 to 100 thousand homes in the Metairie area alone. And that was just in one of many suburbs of the greater New Orleans area.

I drove to a friend's house that I worked with and found them home pulling out wet carpets and furniture. We were glad to see that each other had made it and asked if we knew about anyone else. We didn't. We talked for a little while then got to a point where there seemed to be nothing more to say. The reality was stunning and left us speechless. A feeling of helplessness washed over me. I wanted to do something, but there was too much to do and too many that needed it. I had much of my own and my family's damages to repair. What do we do next? Where do we go from here? Half a million people must have asked those questions.

I wanted to go one more place before returning home. Two miles away from my Metairie house was Lafrienere Park. I had spent many hours there, especially in the last few months before I moved. I had begun walking the two-mile walking path each evening to start an exercise routine. There was a man-made lake and lots of shelters with benches and tables where I would sit and think. Sunsets there were beautiful. There was a huge population of birds, ducks, a couple of black swans and a family of rabbits. I had tried out my new camera taking black and white pictures, and using a disposable color camera, I would take the same picture in color to compare the difference. My favorite photos were of a foggy morning with the haze on the lake with blurred trees in the background. But the gates to the park were closed. It was being used as a dumping ground for the fallen trees and branches from the neighborhoods. There was a mountain of dead tree branches on the spot where there used to be soccer teams of children playing. A heap of wood chips was stacked nearby. The brown color of the chaos replaced the once inviting greens of solitude. How long before the park would be filled again with the joyful sounds of children? Where are those children now?

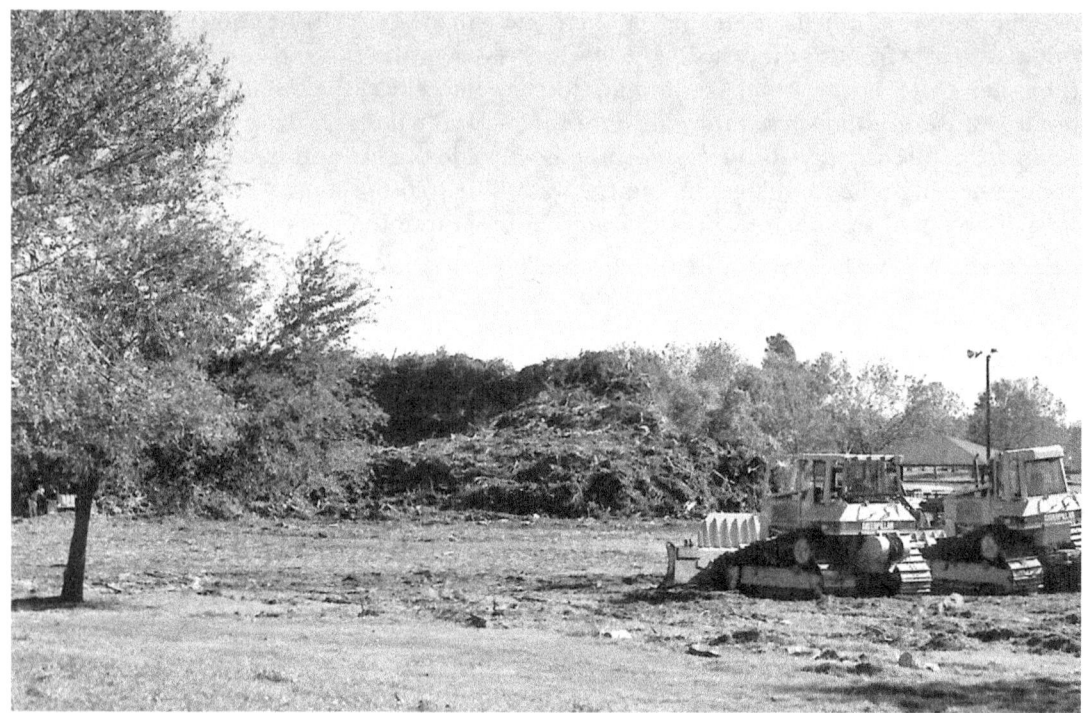

Mountains of debris in Lafreniere Park

Responsibility called as I had to pick up Cherie in Hammond. Rita dropped her off at the dialysis clinic early. They changed her to an earlier time because some of the evacuated patients had returned to their homes or gone elsewhere and they didn't need the late shift anymore. I was supposed to pick her up at eight hence my revelry had to end. The traffic returning was as bad as before, but I got there on time. We ate waffles again and drove back to Rita's house.

Still worried about my aunts and uncle, we located my cousin's house nearby in Mandeville to get phone numbers. We wanted to contact Aunt Lou to tell her of the condition of her Metairie house. She had gone to her son's house in Houston, but I didn't have his number. My aunt was 82-years-old and beginning to show signs of forgetfulness. We didn't know what this kind of news would do to her. We didn't know if she wanted to try to save anything from her house. Even though Aunt Lou's house got less water than Aunt Audrey and Uncle Marion's house near the 17th Street canal, Aunt Lou's treasures were just as ruined. Aunt Lou also had years of family home movies and photos, years of collectables that told of her life story, years of her deceased husband's belongings, years of memories, gone. Whether the water went over the rooftop or half way up didn't matter. What the water didn't ruin, the mold would. Everything they had was gone. Metairie had just opened up for residents, but New Orleans had not. All three of them, my uncle and two aunts, being in their 80's, weren't in any condition to do anything but look at the ruins. By the time my cousins would be able to go to their parents' houses when the authorities approved re-entry, there would not be much left to salvage, if anything.

I had promised Cherie that I would drive her to Baton Rouge on Tuesday to look for her dog. She had left him at an animal hospital in Metairie thinking that he would be safe there. She had heard that the rescued animals were taken to the Louisiana State University Agricultural Center to await the owners to pick them up. We left my cousin's at 1:00 to track down her dog. I found the place easily enough and we went in to see what news they had. The vet took a while before she led us into a private area to tell Cherie that she had bad news. The rescued animals were put in a refrigerated truck and on the six hour trip to Baton Rouge, the truck broke down, the air-conditioning went out and all the animals perished from the heat. Cherie loved her dog and would not accept the news. She thought that they were lying to her and would not leave until they told her the truth. I tried to reason with Cherie, but there was no convincing or consoling her. I knew that she was heartbroken but there was nothing I could do. She vowed to find out the truth before she agreed leave. The story did sound somewhat suspicious, but there was no way to find out what really happened from the people there. Cherie vowed that she would find out the truth no matter what it took. She would find someone who knew what really happened. Until then, she would not give up the search for her dog or the legitimacy of their story.

While in Baton Rouge, I was able to contact Nicolle and have her meet us for lunch. She had her two sons, my grandsons, with her. We hugged and cried. "Everything we had is gone. Our house, all our stuff, the antique car Charles was so proud of!" Nicole cried. "I don't know what we are going to do. I don't know if Charles will have a job."

"Come stay with me." I said, "There's plenty of room. Even though I'm cleaning the first floor, I still have upstairs. We can all sleep there."

"We can't. I have to take care of my in-laws. They have nobody to help them."

"Bring them too." I said.

"They won't leave. We're staying with their cousins where they feel more comfortable. We can stay there as long as we need to."

"Ok, but you know if you change your mind, you can come anytime."

Having seen their total loss, Cherie talked about losing her dog, but knew that Nicolle's loss was much greater. The boys seemed to have accepted their loss better than the adults, but children are more resilient than adults. This would be a story that they would carry with them for the rest of their lives. I could only hope that they would not be affected negatively by it. It was a sad occasion to be reunited like that, but the entire situation was sad. Nicolle had lost her house and all her belongings because when they evacuated, they didn't have time or room to bring much with them. So much sadness and loss and feelings of displacement and helplessness. How were we to get through this?

By Friday, more places were opening. Mama and I went to Covington to wash clothes. Another line to stand in and wait. Another hot and humid day, made all the more unpleasant in the Laundromat. Most of our wash consisted of wet towels and dirty, sweaty work clothes. The other people had stacks and stacks of the same. Everyone had the same expressionless look on their faces reflecting days of waiting and wondering. Some tried to make conversation, but it was all meaningless talk, something to fill the emptiness with.

Later that day, I brought Cherie to dialysis again, but since it was earlier in the day, I went to the Hammond library to check my E-mail again. I don't know what I was looking for, most of my E-mail consisted of Rita and I checking on each other. Since I was with her everyday, there was no need to send messages. I needed to sit in front of a computer and do something "normal." When my time was up, I wandered around the small library looking at the titles of books, then sat down to read the paper. Everything seemed pointless. I aimlessly wandered trying to find something of value to do there. But there was nothing.

When I picked up Cherie, she said that she wasn't hungry. I wasn't ready to go back home being in my aimless mood, so I asked if she wanted to drive around Southeastern's campus. I wanted to see places where I spent my college years. It was a lot different from when I was there, since forty years had passed. There were many new buildings that I didn't recognize, and some that I had my art classes in were no longer there. The campus was still as beautiful as I remembered, even more so. Buildings were far apart and tall pines were dispersed all around. The dorm that I spent my first year in was still there, the same dorm where I heard my first scary story of a man who got off the train nearby and snuck into the dorm and killed all the girls. I think the seniors liked to make that story up to scare the freshmen. It worked. I couldn't sleep that night forty years ago because of that nightmare; now I had trouble sleeping because of a real nightmare. Then I drove by the place I lived when Cherie was born to show her the house she lived in as a baby, but there was no house, just an empty lot. One more thing from the past gone. What was still there was a place called The Brown Door. That's where I drank many a beer with friends instead of studying. It was only a few blocks from the campus just across the railroad tracks. I was surprised that it was still there. I wondered if those new students went there or had their own favorite places to hang out. It didn't look like a place where these new age kids would "chill out" at. I'm sure they had their own favorite places to go to. I had enough reminiscing and so we drove back to Lacombe.

More businesses were opening everyday. By Saturday, we needed to cash checks again for groceries. Hibernia Bank was very well organized and we were out in no time. I was just trying to lighten up the solemn mood that everyone was in by thanking the bank personnel commenting that if FEMA was as well organized as the bank was, we would all be better off. Who was I to complain? I was alive and well, I had a home that was repairable and barbequed hamburgers for dinner.

Sunday, September 11. A sad day to remember another tragic event. So many people were in such a state of somberness after 9-11. No one ever expected to feel that way again, not this soon, on the 4th anniversary. What was happening to our safe, secure world? People are dying everywhere because of Man's stupidity. Is Nature retaliating, or trying to teach us a lesson? Are we destroying ourselves? The future is an uncertainty for thousands of us, not just in the deep south area, but everywhere. What we do here will affect the nation and the world. How we decide to live through this could set an example for the rest of civilization. We can return and rebuild and become just as before, or we can do things differently and show the world that we can change for the better. Our individual plans and routines may have been altered, but that doesn't mean that we can't decide on new and better goals. If our old traditions have been taken away from us, we can start new ones. The homes are destroyed where many of us shared holiday celebrations and family gatherings, but we must find new homes and begin again. What we did and where we did it may change, but not the how. No matter how bad the devastation was we lived through it and must go on. The alternative is to complain, blame others and hope that someone or something will pull us out of this dilemma. There is only one person that can do that, and that is ourselves. Each of us is in control of our own fate. We can question why Nature chose our area to destroy. We can ask why some suffered the worst damage and others had almost none. We can wonder what we did that caused one to die and one to live. We can believe that this was a random act of Nature that had nothing to do with anything, or that we were picked on by a God that wanted to teach us something. But in the long run, we must decide individually what to do next. We must each choose a new path to follow towards our future. This is a time to reflect upon where we want that path to go. Some people will wallow in their sorrow and never change. Others are like the Eagles and will soar above this and become better for it. Some will be vultures and pick at the leftovers ever complaining that there is not enough. Whatever we do, the world will watch, judge and take sides. We have only ourselves to answer to. We have only ourselves to live with.

4.
Recovery

The love bugs have become so plentiful that as I drive on I-12 towards Baton Rouge, they splat on the windshield sounding like raindrops. Where were they during the hurricane? Why did they come back in such abundant amounts? Where were the butterflies, lightening bugs, humming birds, woodpeckers, and red birds? Were they hiding from the 150 plus miles per hour winds, or did they migrate to some protected refuge only to return when they knew it was safe? When the hurricane left, all the animals and insects returned, even the mosquitoes. Nature did not take advantage of this event to eliminate the bad and let only the good return; not the insects, not the foliage, not the animals and not the people. That is why I could not believe in a God who weeds out His creation every now and then at some arbitrary point in time. This was not a biblical Second Act of Noah, as in NOAH (New Orleans After the Hurricane,) but it will probably be recorded as such. Did Noah's God tell Noah which giraffe was worthy of saving and which was not? Were all the animals and people that aberrant that they all had to be destroyed? If I were to believe in that kind of God, then that God is capable of making mistakes and chooses to correct His mistakes by destroying the things that He believed turned out wrong or bad. A "real" God would not make mistakes. But would a "real" God let people suffer, even the good ones? Does He just watch as we endure pain and loss? Or are we all just a product of random acts of Nature, some scientific phenomena that caused us to be what we are? We exist and that's all there is. The only thing that we can do is enjoy the things that make us happy, and make the best of what causes us sorrow and pain.

Mosquitoes cause pain. They were terrible annoyances that interfered with the work I need to do outside. I tried to protect myself from them with sprays and lotions, but they were everywhere and I got bitten anyway. They even interfered with my sleep. Somehow they found a hole in the screens of my open windows and hid inside until I fell asleep. Then they would buzz around my ears, bite me and wake me up. Why do we have mosquitoes? I could find no reason for them, (or roaches), to exist.

Two weeks after the hurricane it seemed that we were no closer to getting electricity. Although there were all sorts of repair trucks on the roads, the orange Asplundh crews that cut trees making room for the electric companies to work, and trucks from power companies from all over the country, but there were just too many of us to restore power to in a timely manner.

My friend, George, finally showed up at my house with a generator. He brought plenty of ice, bottled water, and mosquito spray. He showed me how to start the generator and fill it with gas. He plugged in the water pump so I could have running water to take showers in my own home and clean up easier. The problem was that he could not hook up the hot water heater and the showers were ice cold. Even in ninety degree weather I like warm showers, but the water that comes from the well is from deep underground and is too cold for me. At least things were improving a little at a time and I was grateful to have that much more than yesterday. I could stay home instead of going to Rita's house every night. I had water for showers and for cleaning up. I could plug in a couple of lights, a radio, a fan and anything else I needed, like the essential coffee pot. With a fan I could sleep more comfortably in my own bed in my new home, but who could sleep? The stars were so incredibly bright that I could not stop staring at the sky. Since the whole area was without power, there was no interference from city lights to dim my view of the sky. I would spend many nights looking for the constellations that my father showed me when I was little girl.

There used to be plenty of lightning bugs around my parent's house when I was young. I could catch them in a jar and watch them light up. On my new property, I saw only a few. I would see a greenish glow, but when I focused on it, it would appear in another place, not always in a straight line, but up or down in an unexpected place. Then I would watch until it was gone. I would hear crickets or frogs calling out. There was a buzzing sound nearby that I thought was a bee because it flew around so fast, but when it stopped right in front of me I saw that it was a hummingbird. When I recognized what it was, it flew away. I watched a woodpecker making a hole to find bugs. I think I even saw an eagle, but I couldn't be sure. It was a large bird that flew high up then landed at the far end of my property. Even as I was enjoying the pleasures of Nature that had returned, I knew that I still had my own responsibilities.

By Monday, Rita's electricity was restored. I could wash clothes there and she could start cooking again, except that her stove was making a loud beeping sound. We tried pressing the buttons, but could not get it to work. Eventually, the beeping became so annoying that she had to turn off the power to it. So much for cooking.

A few more things were getting back to normal making life a little easier, but there was still so much that needed to be done. The cable company had not restored service yet, but we didn't have to go fifty miles to Baton Rouge's library to check our E-mail since local libraries had opened. Regular mail was still not being delivered but people could go to the post office to collect mail. Alyssa could not return to UNO for classes since they were flooded and would not be open until spring semester, maybe. She had to go to LSU eighty miles away to take classes if she wanted to graduate on time. She would have to make that drive every day because there was no place to stay in Baton Rouge. On Thursday, Mama and I drove her to class so we could meet with Nicolle for lunch. Nicolle was still crying and depressed. She had so many decisions to make that she could not handle it all. What would they do? Where would they live? Was Charles going to be able to go back to work? What schools would she put little Charles and Nicolas in? What does her house look like? When will they be allowed to see it? Being together helped us vent our frustrations, if only for a short while, and try to concentrate on ideas for the future.

On Friday, I needed to check my bank account to see if I was getting a pay check. I could not continue to cash checks if there was no money coming in. The Slidell library that was closer to me had just opened up. I drove there in the first rain since the hurricane. Although everything was wet, it looked much the same as two weeks ago when I was trying to find a hospital for Cherie. Slidell was closer to where the eye had passed and was particularly hard hit. Much of the city was inundated with water from the storm surge. They were not recovering as quickly as Mandeville and Lacombe since they were closer to the Gulf. Even the mall that was only five miles from my house had just a few businesses open. The WalMart had limited hours and would let a few people in at a time. There were soldiers everywhere, on the streets directing traffic and in the stores watching everyone. They were dressed in their camouflaged outfits and carried rifles. I felt some comfort at their being there, but at the same time fear, wondering why we needed to be so heavily guarded. We were suburban people just trying to live day by day. Were we now each other's enemies when weeks ago we were each other's saviors? How quickly distrust sets in. All I needed was a little more food and water. All I wanted was a few more candles for the night since I still had no electricity, and was only running the generator a few hours at a time. Candles were cheaper than the gas it took to keep the generator going for a night. The moon was so bright many nights that I sometimes didn't even need candles, but somehow they comforted me with their tiny light in the darkness.

Rita came over the next day to help move our books and other items to the second floor. While emptying the kitchen cabinets, I found some dishes that looked ok, but as I removed them, water spilled out. The water must have been in them for more than two weeks now and was dirty and smelly. I rinsed them out and Rita took them home to run them through the dishwasher to sterilize them. They came out clean, but I'm not sure that I ever wanted to eat off them again. Later that day, Cherie wanted to go to the food stamp line to apply for the disaster assistance we had heard about on the radio. I also got some food stamp allowance.

The authorities finally allowed people to go into St. Bernard Parish for their first look at their homes since the hurricane. Nicolle called me Sunday to tell about the horrors that they saw. She, Charles and his brother first went to her in-laws house to see the damage and determine if there was anything salvageable. All three families had houses that were in the flooded area. Because it was three weeks since the hurricane, and no one was allowed to go in to clean up before, the damage had multiplied. They were sickened by what they saw. The reports and pictures on TV were nothing compare to the reality of being there. The stench and ghostly atmosphere couldn't be experienced from second hand reports. Everything was covered with brown mud as if someone had dug out the bottom of a river and dumped it on the city. In effect, that was exactly what had happened. The power of the rushing water brought up the muddy bottoms and deposited it on entire cities. The water was up to the rooftops entering each and every house, then slowly seeped out leaving the mud behind. Everything inside the houses floated or was moved, then settled by the doors as the water drained out. Entrances were blocked by furniture, so they had to break down the doors to get in. Wooden furniture had swelled up, then broke apart as it dried. Mattresses and sofas were so water logged that they were too heavy to be moved. It took both men struggling to drag the wet items out of the house. The color on photographs was washed away leaving only ghostly images on paper. Wooden frames and plaques crumbled in their hands. They had to wear gloves, boots and something to cover their clothes. These all turned black in no time from the filth. When Nicolle finished, she discarded the coverings and even the clothes she wore underneath. Nearby, in a neighbor's bushes was a dead dog that someone had left behind. Cars had floated everywhere landing in backyards, on other cars, on rooftops, and even falling into houses. Even houses had floated away crashing into other houses like an accident on a highway. Houses, cars, everything was covered in the same dull brownish muck. Not a hint of color anywhere but that desolate brown, like a western movie in the dessert devoid of any hue except various shades of brown. Nicolle, Charles, and his brother were so despondent by what they saw, and they didn't have time to go to see their own homes. There was a curfew and no one would be allowed to stay after dark. Nicolle's brother-in-law had already seen her house, since he worked in the area and said that it was the same scenario there as in the whole parish. Nicolle would have to wait for another day, then experience the same misery at her own home. What help could I offer? Nothing except to listen to her sorrow and feel for her vulnerability.

I had to do something to feel useful, and get my mind off the sadness. I cleaned more of the driveway and moved trash and ruined debris out of my house. My hair was getting long and was falling into my face as I worked. It was becoming so irritating that I couldn't do anything without constantly having to brush it out of my face, along with the sweat. The sweat would drip onto the ground as I worked. I knew what it meant by having my blood, sweat and tears going into the land. I was bleeding from mosquito bites and scratches. I cried a lot. Sometimes I would get so frustrated with the amount of work that needed to be done that I could not do anything. I would wander around taking stock of the enormity of the work and get nothing done. The days and nights were still so hot that I would go to Rita's just to get away from it all. We began watching episodes of a canceled show called "Firefly." Alyssa had the DVD's of this sci-fi series. Watching this was another way of mentally getting away from reality. Rita's cable was still not working so we could not watch TV, but could play movies that she had or could rent. Not being able to watch was probably a blessing, since they were constantly airing the aftermath of the hurricane. We didn't need to be reminded of it repeatedly. All we had to do was to look outside.

Monday I got a cheap haircut because I couldn't stand it in my face anymore. My usual place in Metairie was not opened yet. I could drive to Metairie because the Causeway was opened to regular traffic, but there were no businesses opened. My house in Metairie was the only place I could go. It was much shorter to take the Causeway from Rita's house to my Metairie house instead of going around the lake adding seventy miles to the trip, so I could go there more often. This time I remembered to bring my key and went inside. The only damage was in the refrigerator from food that spoiled while the power was off. Since I'm not much of a cook, I didn't have anything like meat that would have spoiled, so that job was not too bad. The refrigerator was saved, unlike most everyone else's. Refrigerators were on the curb in front of practically every house in Metairie, along with furniture, carpets, insulation, sheetrock, toys and unrecognizable things. People's entire lives were piled into heaps on the curbs. Much of Metairie had two to three feet of water or more in some houses, but because we were not allowed to go in for weeks, as in the rest of the flooded areas of the city, the mold was able to grow and spread in the humid climate. Maybe we could have saved more if we had got in sooner, but workers had to clear the streets of fallen trees before it was safe to go in. The trees and branches that were cleared away were brought to Lafreniere Park a mile from my house. There were mountains of tree debris there. Some of it was being made into wood chips to reduce the volume, but even that produced mountains. I drove along Veterans Blvd. on my way back to the Causeway and was astounded by the damage. Especially shocking was the gigantic billboard pole that was cracked in half and had landed on a building nearby. No one could have survived if they were trapped in there. The immense energy that it took to crack a metal pole of that size was frightening. The fact that it was wind and rain that caused it and not a bomb was terrifying. The fact that Man can do as much damage as Nature was sobering. There was enough suffering caused by Nature, why do we feel it necessary to add to human misery by hurting each other as nations, countries, cities, neighbors, friends and lovers do.

Everyone back at Rita's seemed to be annoyed or depressed. Nicolle was still sad when I called to ask if she was going to see her house. Things were happening to improve our lives each day, but happening so slowly that all we could do was wait, wait, wait for the next thing. I wondered if we were beginning to experience PTSD (post traumatic stress disorder.) All the callers to the radio station ask when is FEMA going to do this, when is the Red Cross going to do that, when is the city going to help us, when, when, when. I went to my Lacombe house to be alone and work on things that I could do, like writing about these events, or trying to paint. I spent the next day doing the same. Watching the stars at night helped me stay peaceful and calm. A few times I saw a star that seemed to be moving across the sky. It didn't look like a plane because it wasn't blinking, it wasn't a falling star because it didn't fade out and disappear, and I didn't think I was seeing a UFO. I thought it might be a satellite since it was so far away and moving at a steady speed in the same direction. I was amazed by the things I was noticing since I was not distracted by TV.

Nicolle was supposed to go to her house in St. Bernard for the first time, but the officials closed off all of the parishes again because of a new hurricane in the Gulf. This one was building up to be as dangerous as Katrina and heading towards us. How could this get any worse? Odd that its name was Rita, the same as my sister. Evacuations were started again as the storm built up to a category 5. Another category 5? I could not believe what I was hearing! Was Nature not done with us? Wasn't it enough that our homes and lives were destroyed? Was Nature going to blow or wash away what little was left from Hurricane Katrina? Did we do this to ourselves? Did we cause Global Warming? This time, I seriously considered leaving, but still waited until the last minute to see what this hurricane was going to do. Apparently, I didn't learn anything from Katrina. Once more I thought that since I didn't get too much water from Katrina, and trees didn't knock down my house, I had that false sense of security. I guess some people never learn.

Nothing was normal anymore. Would we ever be able to feel normal again? What is normal, anyway? I spent the day bringing some things to Nicolle in Baton Rouge, and saw where she was staying. It was a nice house that belonged to Charles' cousin's sons who are students at LSU. Eight people were staying in a three bedroom house. No matter how nice it was, it was not home. It would be a long time before she and her family could feel at home somewhere else, along with a million other people. They were all asking the same question. Should we return and rebuild, or start over somewhere else? Where else? Where are the jobs and affordable homes? Few, if any, would have enough finances to find anything like what they had before the storm. Insurance money and FEMA money would not be enough because most people were under-insured. Was that our fault? Most people didn't trust insurance companies anyway. Even with full coverage, the companies find excuses not to pay for complete damages. Anyone who ever had an auto accident knew this. The ordinary citizen usually loses, the insurance companies usually win. This was the mind-set that most of us had and the reason that we didn't have and couldn't afford full coverage. When we added up all that we paid for the different kinds of insurances that we were told we needed, that added up to a big chunk of our take-home pay. There was mandatory auto insurance, mandatory home owner's insurance and flood insurance. I was required to have flood insurance on my Metairie house, but had no flood damage. Nicolle was not required to have flood insurance, but her house had water up to the roof. Go figure! Then there was hospitalization insurance that we felt neglectful to our families if we didn't have it. Plus cancer insurance, dreaded disease insurance, insurance if we lose our job so that we can receive a regular salary if we are unable to work to support our families. Then there was life insurance, not only for ourselves, but for each member of our family. We spent more money on insurance policies than we did on income taxes. I didn't support government take-over of insurance companies, but there has to be a better way. Insurances that were required by law such as auto insurance, and by banks for financed autos and homes, should at least be regulated so as to be affordable to the general public. It is often that extra expense that makes some purchases unaffordable for many. Other countries, (Canada, France) have medical care for everyone. In this country, too many families with sick children and adults have to lose everything just to stay healthy. Their entire lives and finances are so devoted to the ill member of the family that there was no normal life left for them. Elderly people had to decide whether spend their little income on food or medicine to stay alive, but for what? To go through the same decision month after month? Why can't we figure out how to take care of our own? The money wasted in this country on frivolousness is disheartening. The people who make those decisions are demoralizing. They blamed it on us saying that we controlled these decisions by our vote. What are our choices at the poll? Usually it is between one puppet or another and there isn't really a choice. We are told to trust the government, the politicians, and the insurance companies, that they will be there to help us in our moments of need, but they constantly and consistently let us down. The irony of it all is that we really do not have a choice. Our lives and our finances are controlled by the government and insurance companies, much like our spiritualism is controlled by religions. Sometimes the liars and cheaters got caught and punished (somewhat), but we never get back what we lost. We never get back our trust or our money. Is it any wonder that we, the victims of Katrina were skeptical, cynical and confused? We were left in a sea of debris and didn't know which shore to swim to for safety. Our only fault was trust. Our only hope is trust. Our only comfort is our neighbors in this country and around the world who reach out to us with their concern by sending us needed items and help. Our only savior is ourselves. Hurricane Rita began to weaken and started to head more towards Texas. My big sigh of relief could have blown down a tree by itself. But more lives were devastated. More coast was destroyed. More help needed. Would there be enough help for these other people? Would anyone ever want to rebuild, or would our way of life be gone forever?

On Friday, September 23, Hurricane Rita was due south of New Orleans. The probability of it making a sharp turn toward us was slim, therefore we could relax a bit. Still the knowledge of someone else having to go through what we went through wasn't very comforting. Many people had evacuated to Texas and it was predicted that Houston would be hit as hard as we were. Now, they, Louisiana's evacuees, the rest of Louisiana's coast, and most of Houston jammed the interstates and highways to get out of the way. In a way, it was like reliving our horror, this time watching it on TV the way we couldn't watch our own. Sometime around 5 AM., Hurricane Rita hit land. We experienced some strong winds, but not much rain. I feared that the trees on my property that were leaning from Katrina would finally be blown down. Luckily, that didn't happen. There was not much damage here, but the western coast of Louisiana suffered major wind and flood damage. Instead of hitting Houston, Hurricane Rita devastated more of the Louisiana coast. The parishes in the western part of Louisiana did not escape the brutality of Hurricane Rita as they did for Hurricane Katrina. Even without being hit, Houston and its area suffered the effects due to the evacuations having to spend hours in traffic, finding someplace to stay, waiting, then returning. Days were lost. Businesses were shut down. And the Louisiana evacuees there spent more horrific days waiting out another hurricane horror.

Nature seemed intent on destroying our coast and our people. Politicians seemed intent on rebuilding our coast and our city. Even as more of our land was being washed away, the reporters and civic leaders flooded the media with plans to rebuild bigger and better. Even as the city was closed again because of Rita, they want people to return. There were no houses, no jobs, no schools, no supplies, no utilities, but they were planning camper cities in the middle of nothing. This was not the same as getting back on a horse once you fell off. There was no horse. People needed time to rebuild their strength and their characters first. We couldn't simply build a new house and fill it with new possessions without building up our humanity. We experienced tremendous stress when a simple sip of water was the most important goal for the day, when finding a dry place to rest seemed impossible, when the hopelessness of ever finding our loved ones caused unbelievable anxiety, when lighting a few candles in the darkness gave us some measure of comfort and hope. Did our leaders think that talk of rebuilding the city would ease our pain? That's not what we needed. What we needed was time. What we needed was to not be pushed into making foolish decisions while we were at an emotional low. We needed to stop listening to outside interferences such as TV, radio and newspapers, so that we could hear our own inner selves. We needed to make our own decisions based upon what would be best for ourselves and our families. Promises of a bigger, better and safer city were not appropriate. Didn't the politicians understand that we were afraid? Where was their compassion? We wanted answers for our current problems, like getting food and shelter, not promises for a distant unknown future. Two hurricanes wiped away most of what we had in less than a month. Our entire coast was devastated. Would there be another one? We wanted some sense of security, but that was not part of anyone's ability to provide.

When we got a chance to watch TV, all we saw were reports on who was blaming who, and who was suing who. Then they talked of rebuilding, but who would finance it? When people from around the country came to help, there were complaints that these were not locals, and we needed to give our locals those jobs. But there was too much work and not enough locals to do the job. Instead of being thankful for the help, people complained. No wonder tempers were up. We were stressed enough already without the media adding to it.

Natural disasters and human disasters have always had a negative effect on our lives. People died and treasures were destroyed. Bomb it, then rebuild it. Nature destroyed it? Then rebuild it. What's the point? The message seemed to be that possessions were more important than people, that memories were more important than morals. If we fixed it like it was, then we will be "normal" again. Was getting back to normal more important than raising the norm to be a more caring and humane place us all to live together? We would all come out of this differently, but would we be better? Many will continue to suffer and complain, "Woe is me. I lost, I suffer. I, I, I, me, me, me." I am reminded of a great man who once said, "Ask not what your country can do for you, but what you can do for your country." (John F. Kennedy.) Perhaps we need to revise his statement to read, "Ask not what you can do for me, but what I can do for you." It is in losing ourselves in other's suffering that we forget about our own. Or as another great man said, "Give and you shall receive." (Jesus.)

September 23 was also Cherie's birthday. Months ago, we planned a party for her at my house, but I could not have a party because most people would not be able to come, I had no kitchen, and besides, Cherie was angry and would not come anyway. She spent most of her life being angry at her condition and "the system" for not helping her. She said that we should be able to understand how she feels now that we are all in the same boat. Confused, lost, alone, and stranded. Rita and I had cake anyway and would give her some when we saw her.

I had been trying to get Nicolle and her family to come to Lacombe just to get away for a day or so. She finally accepted. I offered to pick her up and bring her back later. Hurricane Rita had knocked the power out in Baton Rouge. Nature seemed to want to continue to remind us not to let up our guard. All over Baton Rouge there was another shortage of gas and ice. I brought her a big ice chest and plenty of bags of ice to help out the eight people there, but by the time I got there, the power was restored. But not for long. After we left it went out again. At least the ice I brought would help save some of their food from spoiling and causing another unnecessary expense. I took Nicolle, little Charles, and Nicolas back to Mandeville where we stopped at Rita's for lunch. Rita and I had gathered as many photos as we could find to replace the ones she lost. We put them together and gave her the pictures.

Then we went to Lacombe and discussed some of her options. We had heard that FEMA would supply trailers to victims and she wondered if she would qualify. FEMA would set up trailer parks, but would also give a trailer to someone who had a place to put it on. These were unqualified rumors and everyone began to wonder if even FEMA knew what they were doing. Since I had seven acres, there were many places to put a trailer on my property. We looked around my property to see if there was a good place for a trailer. There were quite a few possibilities. Her other options were that Charles' family also had a large piece of property in Picayune, Mississippi, and offered her and her in-laws a place there. She could even move in with me, but she felt that she had to help her in-laws and could not leave them. I said that it would be ok for them to live here too, but she said that her in-laws would be uncomfortable, and besides my first floor was still not ready to live in. They would not have to hurry to make any decisions anyway, since no one knew what was going to happen about anything. Four weeks had gone by since Katrina and no one was any closer to solutions about anything. A million lives put on hold. A million lives with no plans, no security, no idea what to do, where to live, where to work, who would help, or when that help would come and in what form. A million temporary homeless people waited in line for something, not even knowing what the line was for. The line did not seem to be moving.

When I brought Nicolle back to Baton Rouge, she said that it was the best day she had since before Katrina. The little things that we do for each other really do help.

Driving around in total darkness was strange. The interstate was not that different as there were lights from other cars, but when I would take the exit to Rita's or my house, there was nothing to see that would tell me where I was. The landmarks that I was getting to know were hidden in the shadows of the night. Street lights and business signs were gone. Without electricity, there was nothing to light up my way, and many other signs were blown away. There were only the mysterious black shapes of trees and buildings against the dark blue of the night sky. Many times I passed up my house because of the unfamiliarity of the landscape.

When I arrived home, Cherie was waiting in the driveway. She wanted to spend the night with me to discuss the problems she was having. Her social security check was lost in the mail. No one was getting any mail. We could not check on bill payments or anything. My land line began working intermittently, but whoever we tried to call, we got busy signals or no answers. Cell phones sometimes worked, but calls would be dropped in the middle of conversations. Most New Orleans businesses were not even there to receive calls. Even doctors were unavailable for appointments. Cherie was aggravated and frustrated all the next day because we could not get anything done. She argued with me, she argued with Nicolle when she got in touch with her on the cell, and she complained to other people that she was able to call. She was calling people that understood her and her problems. When she got in touch with another friend of hers, she got more bad news. The lady's son had just hung himself a couple of days ago, which was on Cherie's birthday. Saddened by this news, she kept venting her frustrations out on me. I tried not to argue with her but that only riled her up more. I have learned to listen to her and not offer my opinion since she didn't listen to my ideas anyway. But even when I didn't join in on the argument, she would get angry because she would think that I'm not interested. She is a very difficult person to deal with because of her problems, and the situation in the aftermath of the hurricane had only made it worse. At one point, I had to get away for a while, so I went to a nearby store and got us a couple of cold drinks, thinking that if there was no one to argue with, she would calm down. But when I returned, she just picked up where she left off. This continued until 3:00 PM until she left for dialysis. She always left me feeling frustrated because I couldn't help her the way she wanted to be helped. She knew how to push my buttons. Later I went to Rita's to sleep. She got her cable back so we could watch TV and use the internet. But the cable kept going out. It was better to have nothing, than to have something work intermittently, not knowing when it would work again.

The next day I drove to Metairie and visited the school where I taught last year. They had just reopened for the teachers to begin assessing the damage and start fixing up their rooms for the students who would return the next week. They expected about half of the students to return. All of the teachers came back, many with no place to stay except with friends or family since so much of Metairie had damage and most of them lived in Metairie. They wondered if some of them would have to be let go since many students would not return. Another place with questions and no answers. They, like everyone else, would have to wait and see and wonder if they would have a job or a home.

Somehow, the drive across the Causeway always relaxed me. I was not going to listen to the radio to hear people asking the same questions and getting the same vague answers. I preferred to play a CD that was a favorite of mine when I had to drive a long stretch alone. The soundtrack from "The Last of the Mohicans" always relaxed me. I would turn the volume up and not think of anything, just let the sound of the music inside. I looked out at the blue/green water of Lake Pontchartrain. The lake was silent and deep as an occasional pelican flew alongside the bridge. The sun reflected on the water like a glistening silver glaze. The sunrises and sunsets on the lake were extraordinary after the hurricane. It was as if Nature was apologizing to us by giving us exceptional beauty after an angry rage.

Everything seemed to be back to normal then. But it was not normal. There was a difference to this peaceful appearance. There was a feeling that the lake had a hidden agenda. It was as if it were calmly waiting and planning for another attack. It could do what it wanted no matter how high we build the levees, or how much we protected our homes. Funny how the ordinary soothing waters that I enjoyed before now cast an ominous feeling that was never there before, a threatening feeling that she could become a terrifying destructive monster.

As I drove and listened to the music, I began thinking. *There are people like that. They can look so innocent and calm, but suddenly and without warning they can go into a rage just as destructive to others as a natural disaster. Is this the way Nature is? All of Nature, and all of us can go on day to day, living the same routine, and then something clicks, something triggers Nature, animals, people and they become enraged, strike out, and destroy. The beautiful and the innocent get hurt. Then time goes on, wounds heal, and life continues. But the evidence of the rage does not go away. Scars are there on the earth and on the people. Scars are a reminder that we should not trust serenity. It can happen again, and it will.*

Then my thoughts turned to what the most important thing was that I lost. I realized that it was not a physical thing. Physical things could be replaced except for photos, documents and family heirlooms. But as important as these things were, it was the emotional devastation that was the most fragile of all. This was a tremendous blow to our confidence, security, safety and trust. This destroyed our feeling of invincibility, that we could live our lives confident that nothing bad would happen to us. There have been wars and natural disasters before, but they happened to other people, usually small groups of people, or other nations, not something as massive as this, affecting so many people at one time. In the past there have been momentous events that changed our lives, but we had the flexibility to deal with it, accept it and move on. I thought of the first time I, and this nation during my lifetime, was faced with such an incident. I was a student at Southeastern Louisiana College in Hammond, Louisiana when John F. Kennedy was assassinated. The thought that someone who was in such control, who was so young, vibrant, and powerful, could be shot down by an insignificant individual was incredible. Kennedy had all of this power and promise, but he could not control his own fate. It didn't seem right for that to have happened. It was this loss of control over our future that made us conscious of our own vulnerabilities. This was what happened to us after Katrina. On this massive scale, a million people lost control over their futures. I realized then that what made people so frustrated and angry was their loss of control after something like this. When we were promised that help would come in whatever form, we had no choice but to wait for it. We became stuck in a situation that we could not get out of. The people that lost the most were the angriest because they had no place to live, no money and no transportation. They lost their freedom. At least I still had that. At least I still had a place to live, a car and some money to go where I wanted and do what I wanted. Many times in my life when I felt out of control, it was my anger that produced action. Usually, those actions were positive for me, but sometimes those anger driven actions are negative. When I became angry at my abusive ex-husband after realizing that he had total control over my life, that I finally was moved to get a divorce. That turned out to be a positive action on my part. Feeling helpless about my children's lives also produced anger in me. I had no control over Cherie's illness and that made me frustrated, and still does, because there is nothing I can do for her. I had no control over Ricky's addiction and I could not help him. That made me frustrated. I wanted to do something to help Nicolle because her family was trapped in this homeless situation, but there was nothing that I can do for her either. All I could do was listen and let them vent their frustrations and then offer suggestions. Whether or not they take my advice is their decision, and I must accept that as their right. The only thing I had control over was my own life. Somewhat. Although Katrina changed so much of this area and its people, I still had much of my freedom to be grateful for. So many people were angry because they lost their freedom to do as they pleased and live where they wanted, that they had no recourse but to be angry. Being angry, they want to blame someone or something for their situation. They can't vent anger at Nature's storm, so they direct it at the government, and its representatives. I discovered that my anger, although directed at other people, was really anger at myself. I was angry because I realized that my situation in that abusive marriage was my own fault for letting it happen. I was angry at myself because I should have done something sooner, I should have had more control over my life, but I let the bad things happen until it became too much to bear. This anger was a negative thing until I let it change my life for the better. Then it became a positive action. It was a difficult thing to face my own mistakes, but in doing so, I could make peace with my past and control my future. I am not in the position to give advice to anyone else, but if I could it would be this: stop living in the past; stop expecting someone to save you; take control of your own life; find excitement in doing something different like living somewhere else, making new friends, or enjoying the beauty of Nature; help someone else; and stop being angry. If it sounds easy, it's not. It is the hardest thing anyone can do, to change an entire way of life. But Katrina happened and we have to do something. Instead of making this a negative event that we were victims of, turn it into a positive thing that Nature gave us a wake up call, an opportunity, to change and do something different with our lives. Not to be clichéd, but wake up and smell the roses. Turn a negative into a positive. When I had no electricity, I looked at the sky and saw stars that I had not seen before. When I didn't have much food, I looked in the mirror and said that this is a good time to lose some weight. When there was no one to help me move trees and branches, or clean up inside the house, I decided that I needed the exercise and went to work on it. When I knew that assistance was a long way off, I needed to be patient. As long as I had something to eat and a place to sleep, I would be ok. I would use this time to examine my feelings and my life. I would write down my thoughts about all of this, and if nothing else, my children and grandchildren would have a first hand account of it. With no

TV to occupy my boredom, I would let Nature entertain me. I would try to understand what this life was all about, and what my purpose here was. I have heard it said that life happens in the "now." I used to waste too much time complaining about the past, and anticipating, or waiting for the next good thing to happen, but in doing so I forgot to enjoy the present. While writing this, I am enjoying my life because I am doing something that I want to do. No matter what comes of it, it is me, now, being me. I could use this time to think of all the work I have to do yet with cleaning and repairing, and be bitter about the slowness of it all, but I know it will eventually get done. I cannot stop living in the meantime. This cleaning up process is not about getting my life back, this IS my life. This IS the journey. I can enjoy the process of raking leaves, mopping floors, painting walls and picking out new things to replace those that were lost, instead of being resentful that this happened and there is no one to help me. Katrina left us all at a crossroad where we are forced to make new decisions about our futures. Each one of us alone must make a decision to go back to the way we were before, or to change into a new, better person and live a better tomorrow. Each of us can decide to return to the electric lights of the cities so that we can sit in front of the TV and complain, or we can go out into the darkness and look for a falling star and rejoice at its wonder. William Faulkner wrote in *Go Down, Moses*, "...you can't be alive forever, and you always wear out life long before you have exhausted the possibilities of living."

Saturday, October 1. My first visit to Nicolle's house. I stopped at Rita's for breakfast, an omlette cooked in the microwave, (she was still waiting for a new stove, and since her old microwave broke, I gave her my old one.) Rita rode with me to Nicolle's house. We drove to Baton Rouge to pick up Nicolle and Nicolas. Since we were going I-10 through Metairie, we stopped to visit my Aunt Lou. She was there with her son and grandson who were helping to clean out her house. She had about three feet of water and since they were not allowed in sooner, there was mold everywhere. She was as lost and confused as everyone else, not knowing what to do or where to live. Seeing her possessions being piled onto the street was depressing. Conversations were nothing more than talking about how awful it was, where we all went, how we all were, and what to do next. There seemed to be nothing more to say. Any talk of normal things was insignificant and forced. We all had so much to do, that we could not stay for a friendly visit, so we left. We drove back on I-10 through Metairie. As we looked down the streets, we could see more piles of possessions being thrown out. It was when we crossed the parish line into Orleans that the complete devastation from the flood waters was obvious. In Metairie, the water came up to a few feet into the houses, mostly between I-10 and Lake Pontchartrain, and people had been allowed in to clean up. But as soon as we crossed the 17th Street canal, which is the border of Jefferson Parish and Orleans Parish, it was like crossing into the twilight zone. Metairie's devastation was hard enough to comprehend, but this was immensely unbelievable. The odor of rotting wood and other unimaginable things permeated the air. The color of brown/gray was settled on everything, from the houses to the trees and roads. The emptiness circulated ominousity everywhere. A city deserted. Quietness abounded. No people, no animals, vegetation brown and dying, nothing living for miles around. Very few vehicles were on the interstate going in either direction. Along I-10 through New Orleans East were many apartment buildings, windows broken exemplified the emptiness inside. Muddy vehicles were in odd places and positions everywhere as if an angry child threw his toys in disarray. The car lots had muddy cars and trucks still lined up as if waiting to be sold. Why didn't they float around like the other vehicles, I wondered. Passing by Six Flags, there was water pouring out of a small man-made levee surrounding the park. It looked as if it was deliberately cut to drain out the water that was trapped inside the park. As we went up the bridge over the Mississippi River Gulf Outlet there was water where there once was land. Before the hurricane we could see the swamp between the bridge and the New Orleans skyline. Now, the Gulf water had eaten away the land and all we could see was water. Along the highway into St. Bernard Parish, there were boats everywhere. Brown, muddy, and broken, they littered the area. As we passed the military checkpoint into the Parish, everything was brown. The mud had dried on the trees, houses and ground. As it dried, it formed into small geometric cakes, reminding me of a gigantic puzzle waiting to be put together. Along St. Bernard Highway, we could see the black stain of the oil that had leaked out of the refinery. It was settled on the ground, grass, buildings, and piles of debris. As the water and oil receded, it dried into horizontal stripes on the buildings that were still standing. In an area that used to have new campers for sale, the campers were strewn everywhere, floated across the highway into a neighborhood and down the streets. When we got to Nicolle's street, she wanted me to drive down the block to see some of the other destruction first. There were two houses that had been lifted off their slab and floated across the street into other houses. The fence around the property was undamaged. The water was high enough to have floated the houses over the fence without damaging the fence. This same scene was repeated all over the parish. Cars had floated everywhere, on houses, on each other, blocks away from their original locations, and settled wherever they could, as the water receded. Some were in odd precarious positions, sideways, in backyards, on trees, even straight up and down under carports. Even in the most creative of movies of wars and destruction, the writers could not have conceived of this scenario. Charles' car was a block away, almost in someone's back yard, with two other cars on top of it. Every house had a large "X" spray painted in neon colors on the front, on boards on the doors or windows that the occupants had used to cover the windows, or directly on the house if there were no boards. In the spaces of the "X" there were painted letters and numbers that indicated the date that the military was there, the unit that was there and the number of bodies that were found in the house. A chill ran up my spine as I saw this written on house after house after house. It reminded of the scene from the movie, *The Ten Commandments*, where the houses of the Jews had on marks on their doors so that the Angel of Death would pass by and not take their first born sons. Except the Angel of Death did come here. The Angel of Death did take lives and the evidence was marked on the doors of the houses. As much as we didn't want to look, we could not escape the urge to see if there was anything but a zero on any of the houses. On Nicolle's house the markings showed that the date was 9-18, indicating that the military was there on September 18. The unit was EBR, which we thought meant East Baton Rouge, and the number "ZERO," meaning that there were no bodies in her house. The memory of seeing that painted in bright orange spray paint will be in our minds forever, as with everyone who went back to see their houses. It represented the death of a lifestyle. It was the same as the memory of a person who died, seeing them for the last time in their coffin. Like the funeral of my son, Ricky. We wanted to remember the better memories of him, and do, but then the final image of him lying in the coffin always comes back up. We wanted to remember the good days spent in the houses that were destroyed, but the final image of the neon paint on the front of the house will come back to remind us that something ended. A person had died; a home was destroyed. Perhaps that is why some people will never return. They don't want that to be their final memory. Perhaps that is why some people can't attend funerals. They want to have good memories of the loved one, not that of him lying in the coffin.

We talked about the service men that had the job of going into the houses, and wondered about their feelings as they broke their way into each and every house, not knowing what they would find there. They had to go inside and make their way through the slush and scattered muddy furniture into each room to check for bodies. How did they feel when they discovered a body? How did they feel going into house after house seeing the destruction first hand? These were mostly young men, barely into their manhood, for many it would be their first experience as soldiers. This would be an incident that they would have to carry with them for the rest of their lives. I wondered if experiencing a war would have been better for them, or worse. That is how the hurricane affected not only those of us who lived through it, but also the thousands of military personnel, volunteers, reporters, and whoever else was involved with the rescue and clean-up.

The streets were not cleaned up like they were in Metairie. We had to drive over mud and branches, and around trees that had fallen and were still on the roads. When the water receded in Metairie, it did not leave layers of mud. The water in St. Bernard Parish came from a different source and stayed longer. Curiously, from the outside, the houses did not look like they had much damage, except for the color of mud, and the markings on the doors, they were still standing. It wasn't until we went inside Nicolle's house that we could see the strangeness of what had happened there. The mud that came in with the water did not leave as the water seeped out of the houses. Instead it settled on everything inside. After the furniture and other belongings floated around then settled in unusual places, the mud, thick, wet and dark brown, like a child's finger paints, stuck to everything leaving an unrecognizable mess. It was only after scraping off the mud that items could be identified. The mud was inches deep on the floors, still wet and slushy. To go inside the house we had to wear rubber boots up to our knees, and disposable gloves if we were going to touch anything. Nicolle and Charles had gone there before and opened the doors and windows, but even though the days were hot and dry, nearly five weeks later, the mud had still not dried. As we entered the house, the odor was like going into a septic tank. The gooey mess on the floor stuck to our boots making it hard to walk, and as we lifted our feet out of the mud there was a sucking sound as if the mud was trying to hold us in its grip. The refrigerator was lying on its back on the kitchen floor and the sofa blocked the way to the back of the house. These were too heavy to move, so they had broken the windows to get inside the bedrooms. The ceilings had caved in and we could see right up into the attic. The wooden blades on the ceiling fans were covered in mud and drooped down, still attached to the ceiling. The bathtub was filled with mud and who knows what else. There was no way to drain it out. The kitchen cabinets had fallen off the walls, but some of the delicate dishes inside were not broken. Books and photo albums were glued shut. Pictures that were still on the walls were washed out, leaving something that looked like a Dali painting with faces and objects distorted. Nicolle and Charles had been ravaging through this mess to find any precious items that they could save. They had found some jewelry, photo albums, some of their kid's toys that had been saved in a plastic bin along with some of their school work and other paper work that was important. Many other people would come, take one look at their house and leave, taking nothing, deciding that it wasn't worth it. Many had not even bothered to return and never will.

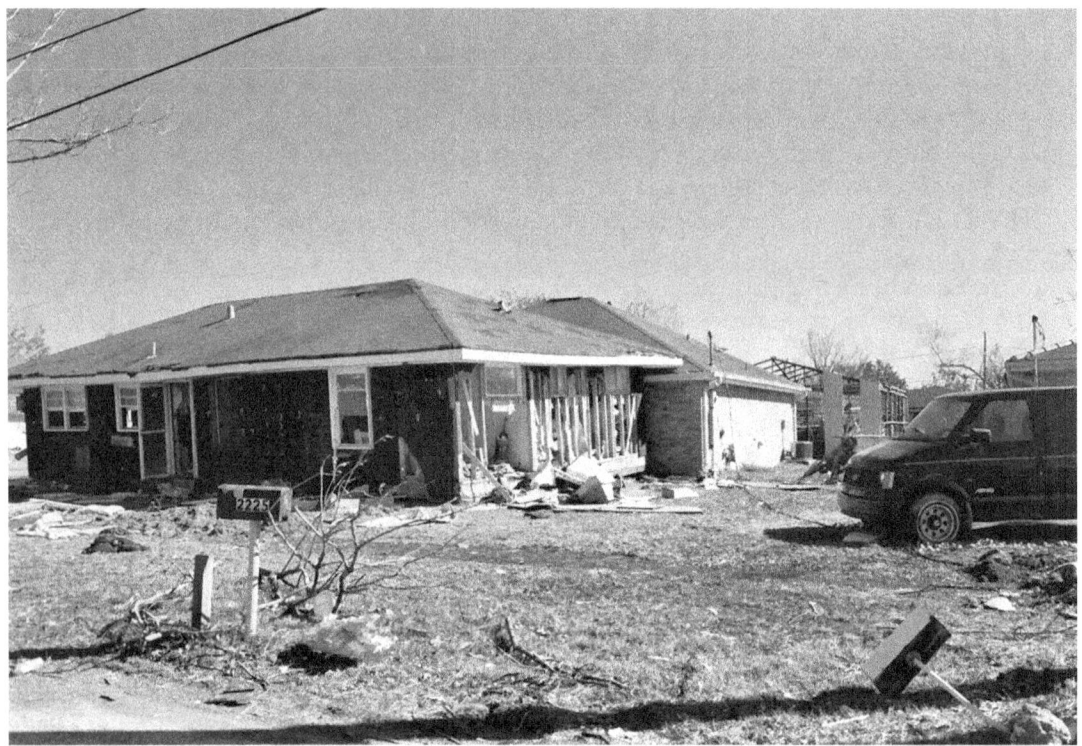

Chalmette house floated into another house

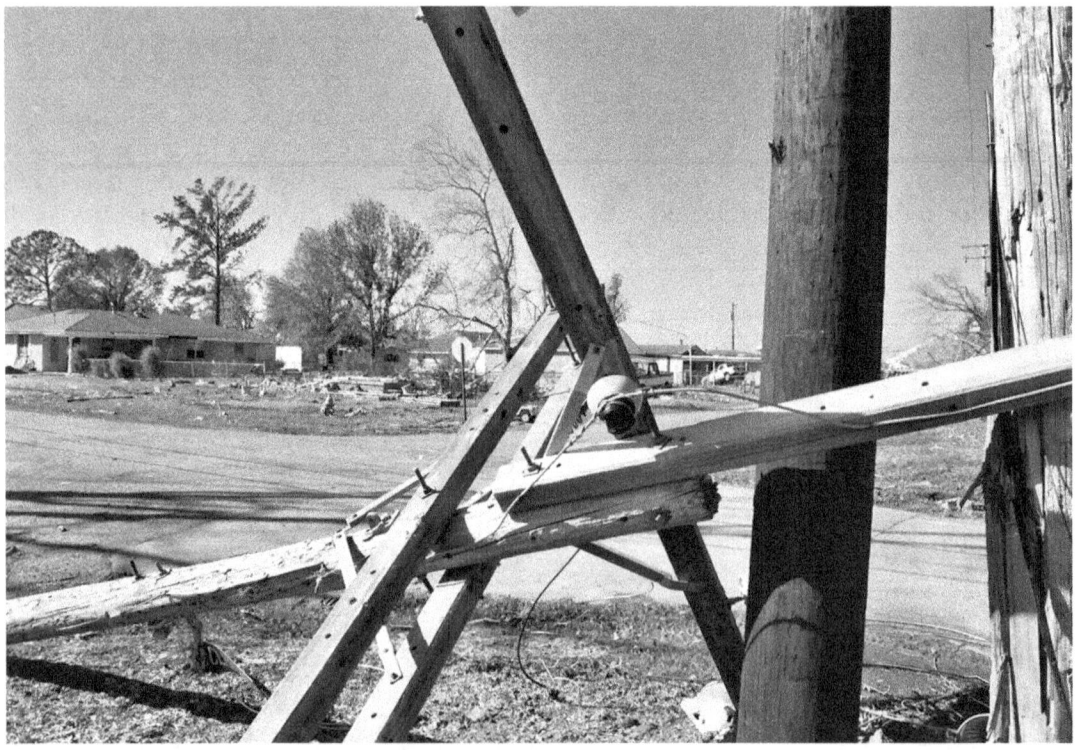

Empty slab where the house used to be

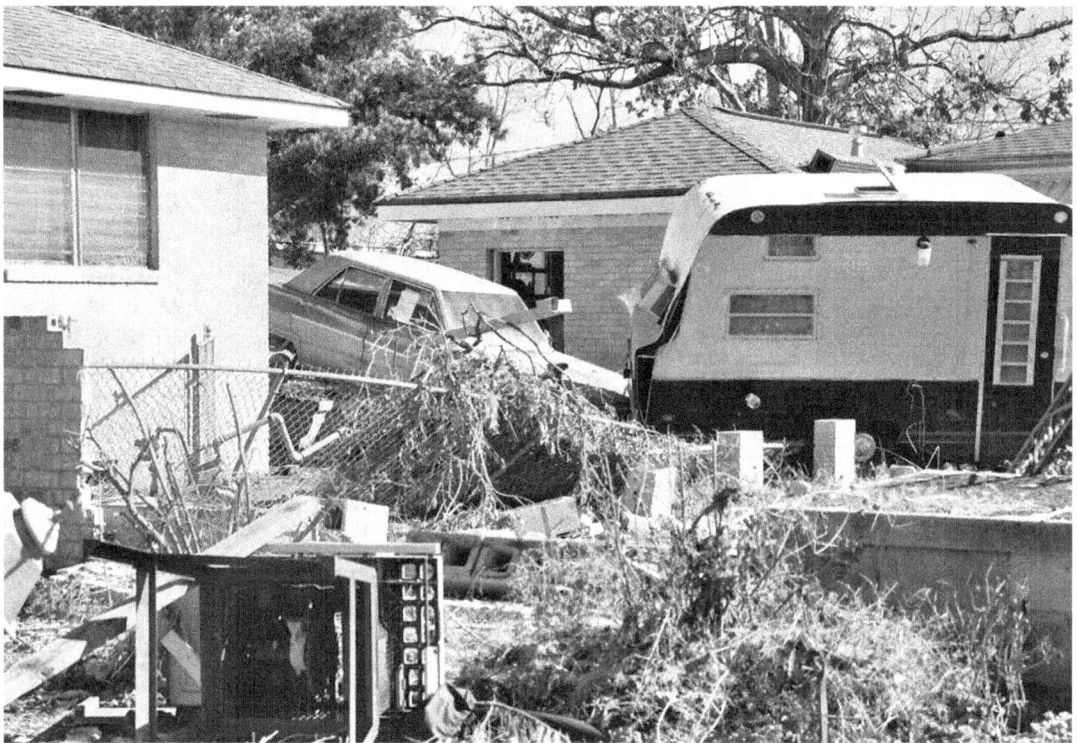

Car floated from blocks away into a back yard

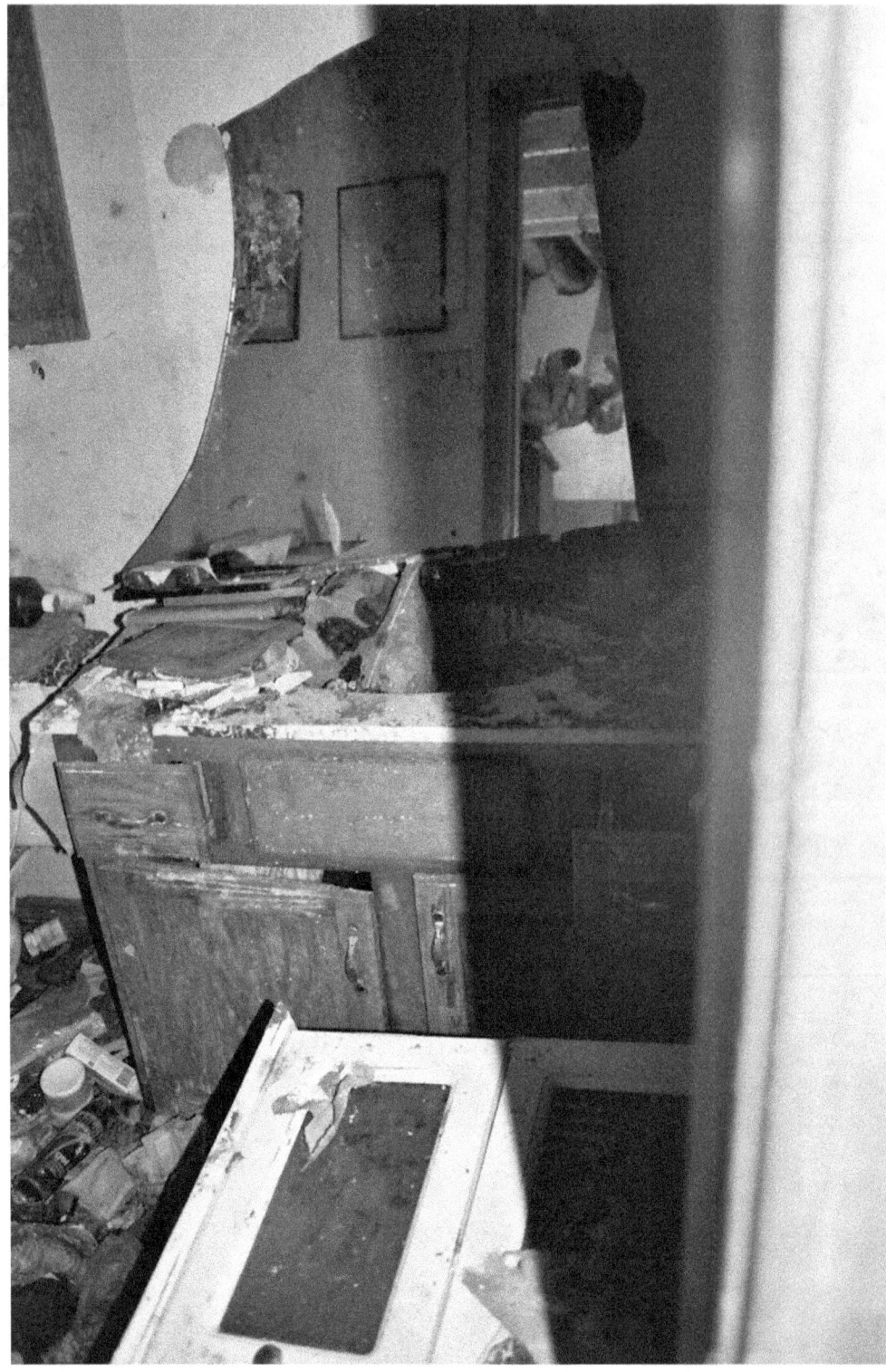

Unrecognizable mess in a bathroom

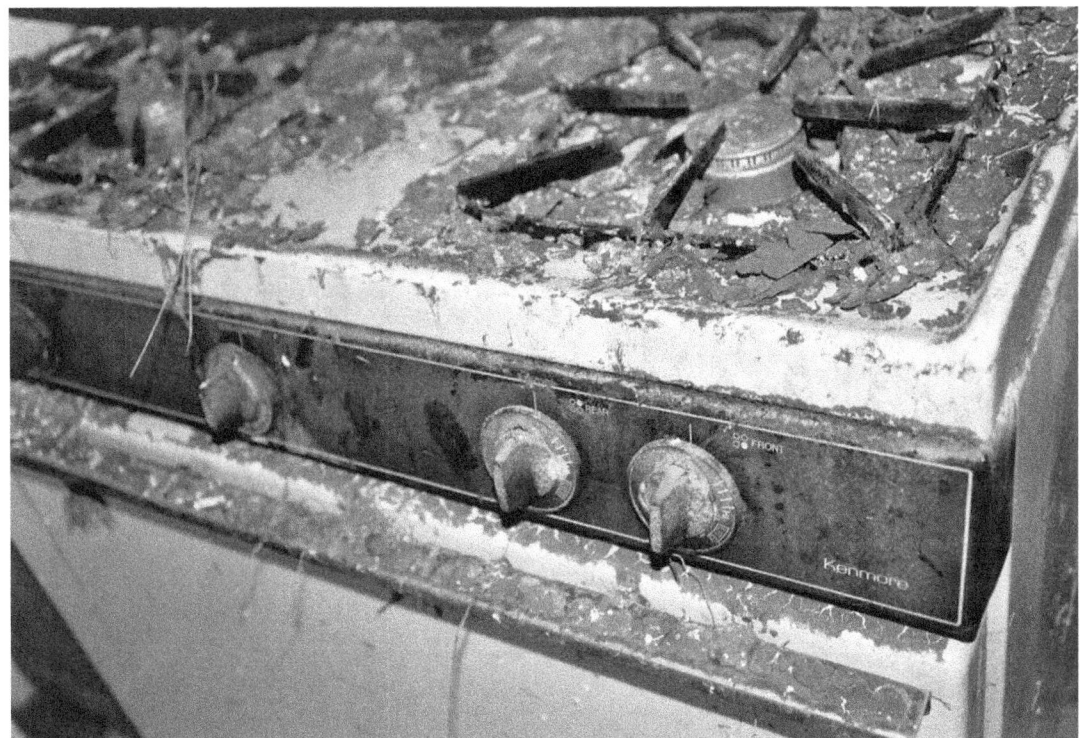

Debris on stove

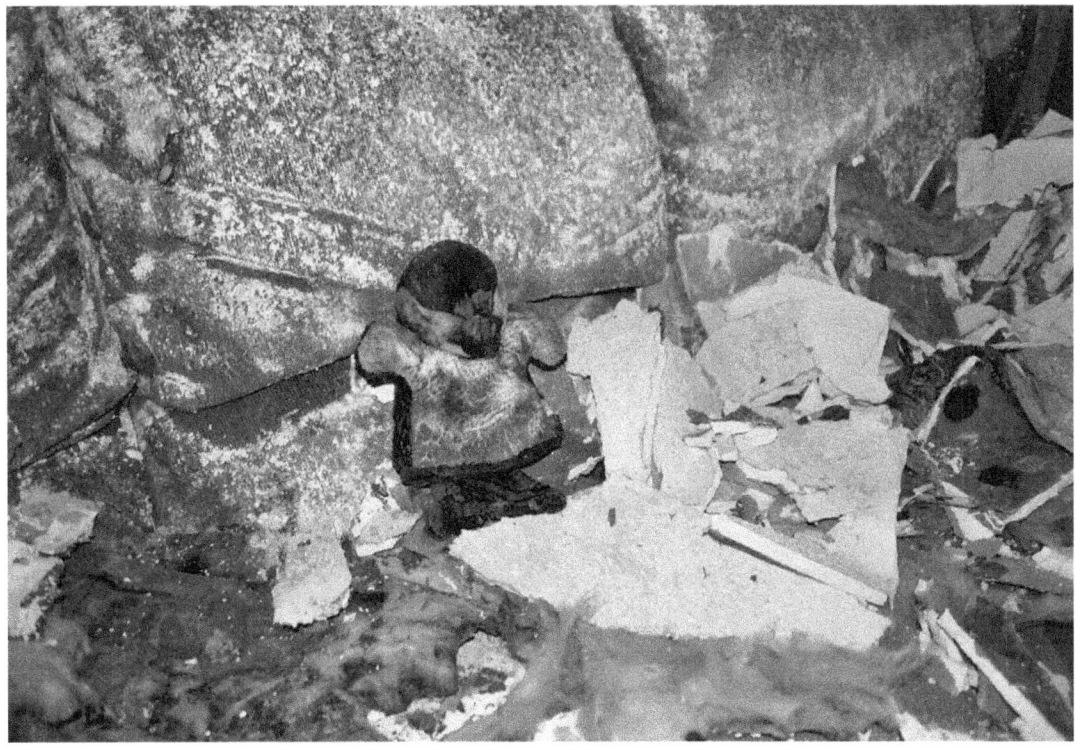

Stuffed animal on sofa

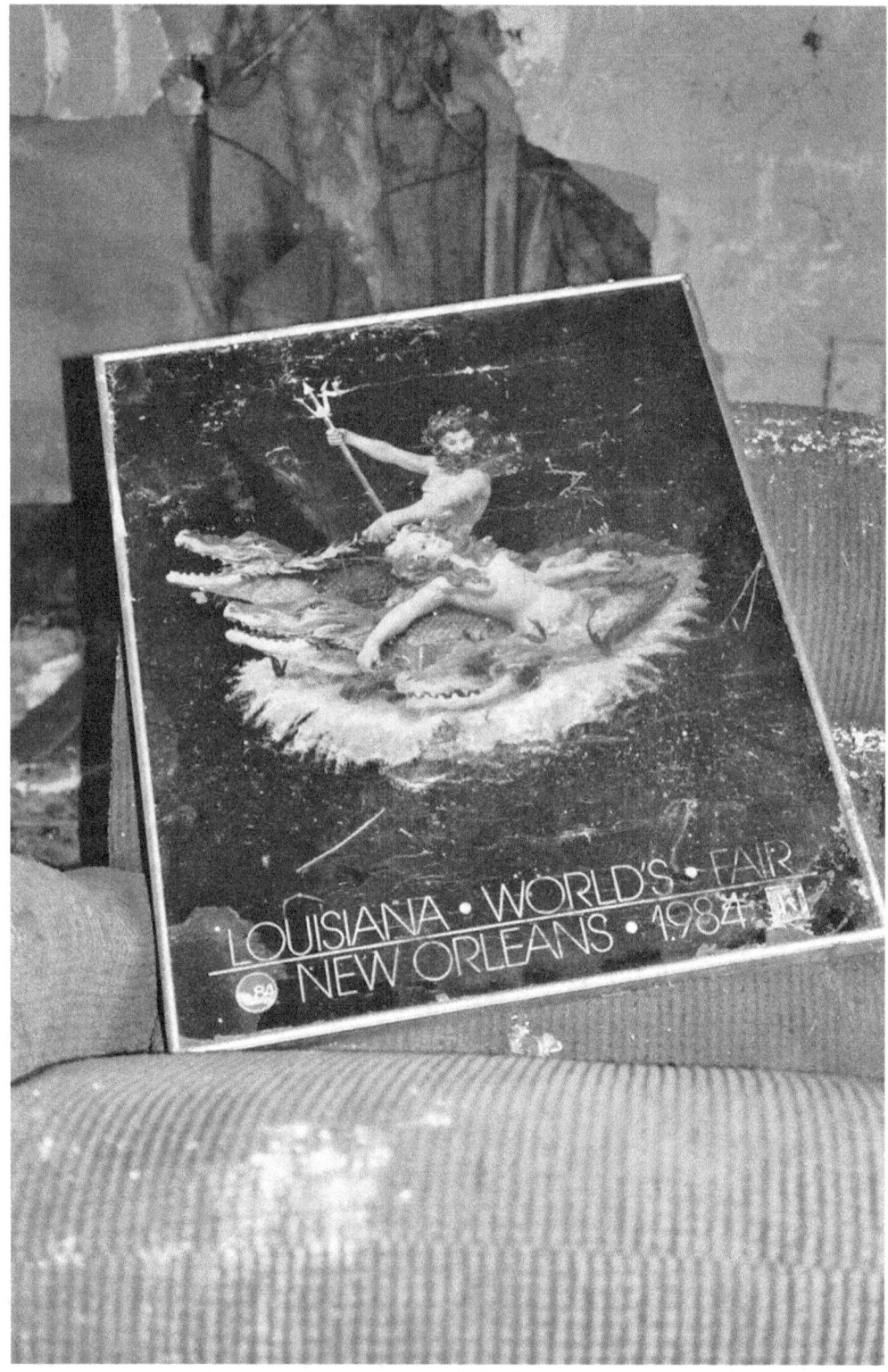

1984 Worls's Fair Poster. The theme was "Fresh Water as the Source of Life"

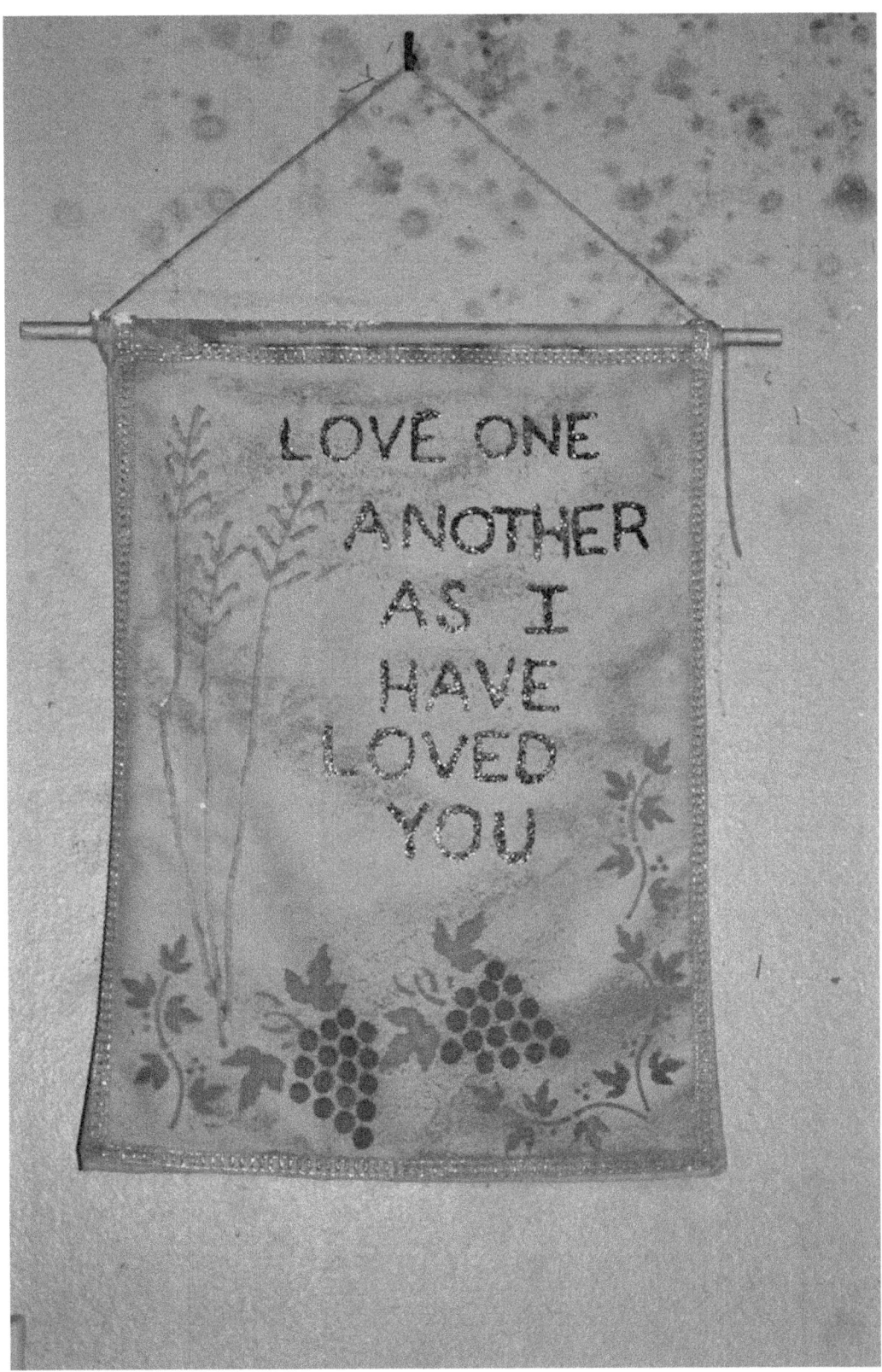

Mold on wall banner

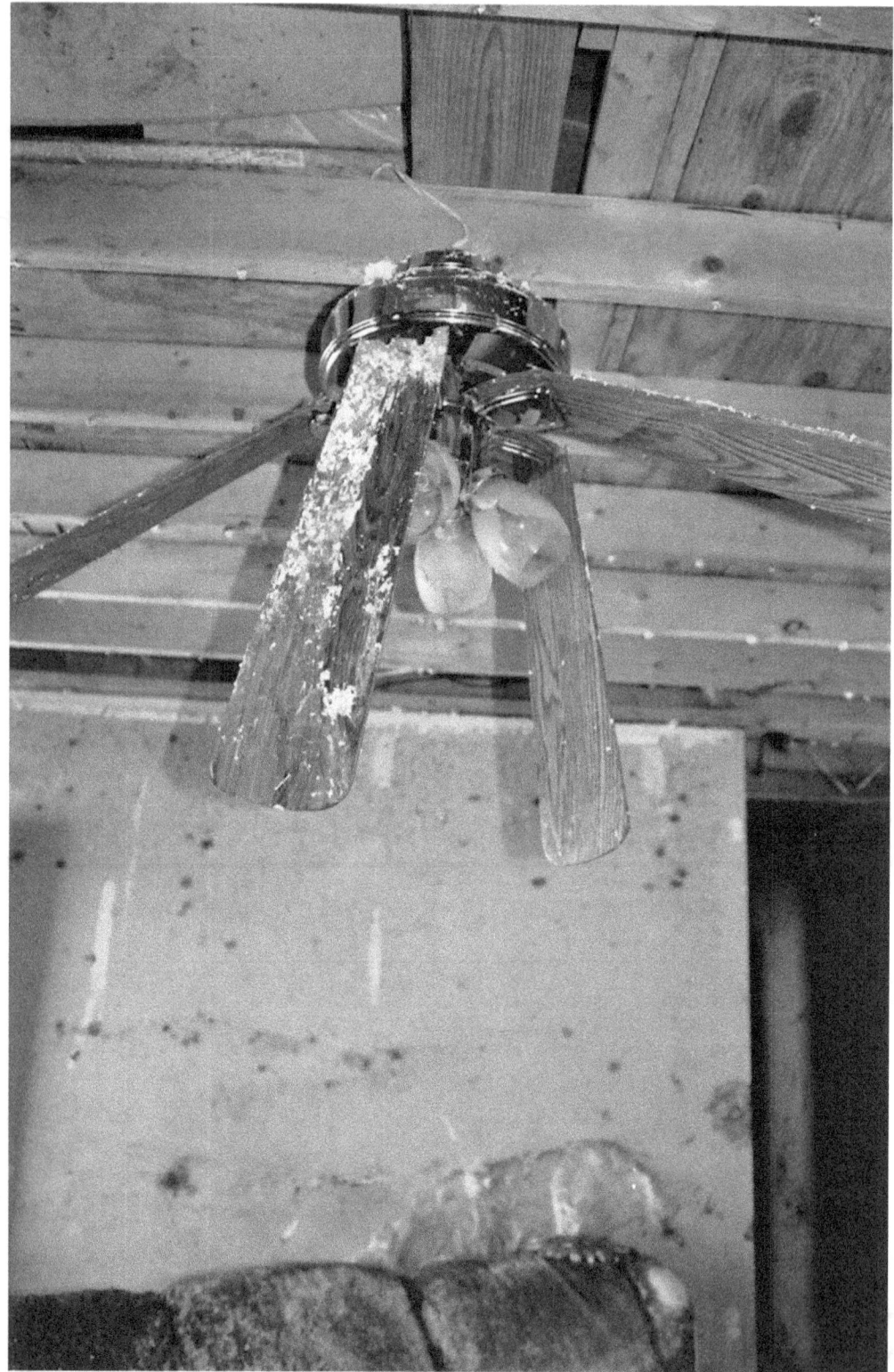

"Dali" ceiling fan

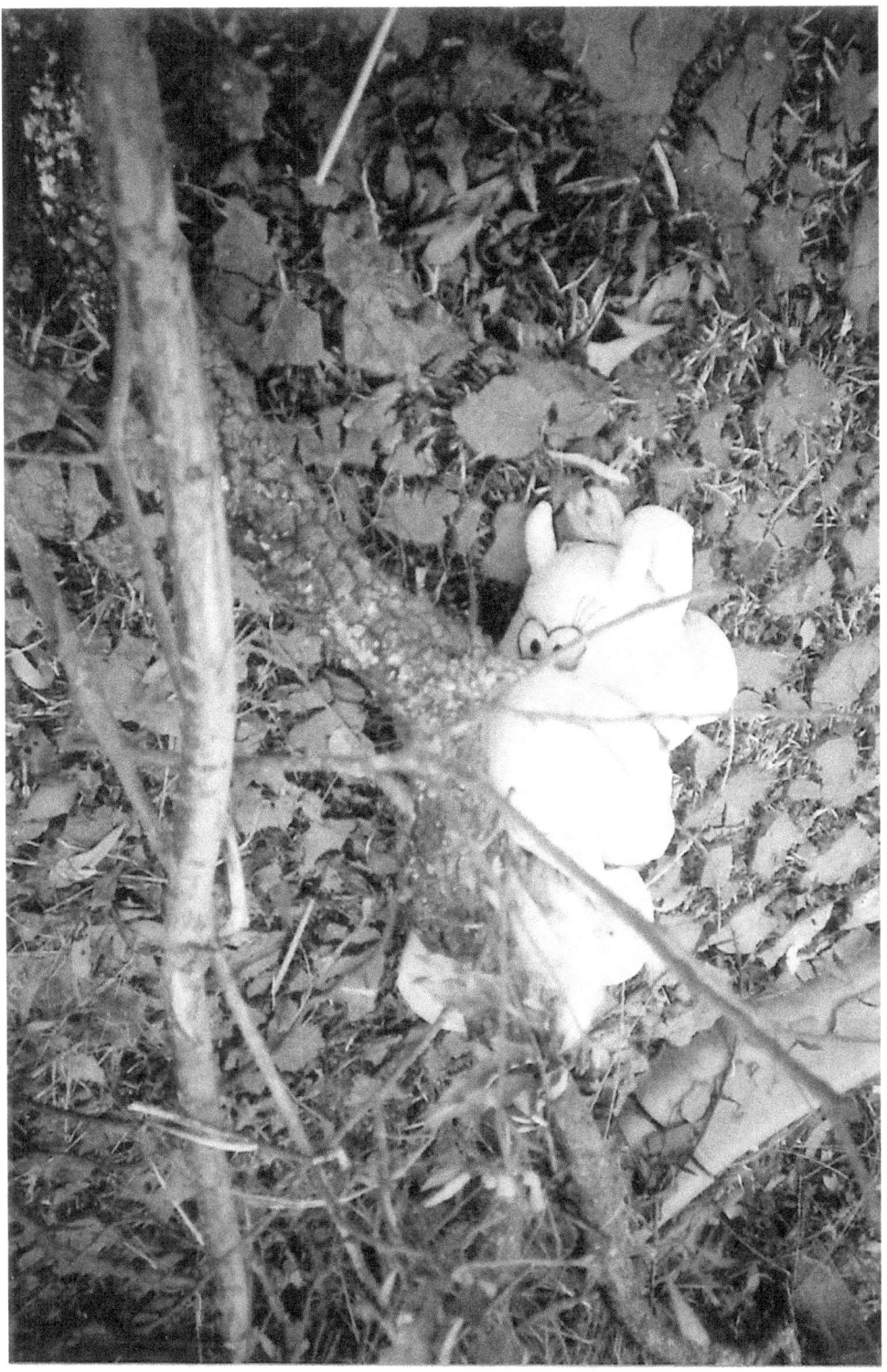

Stuffed animal on mud

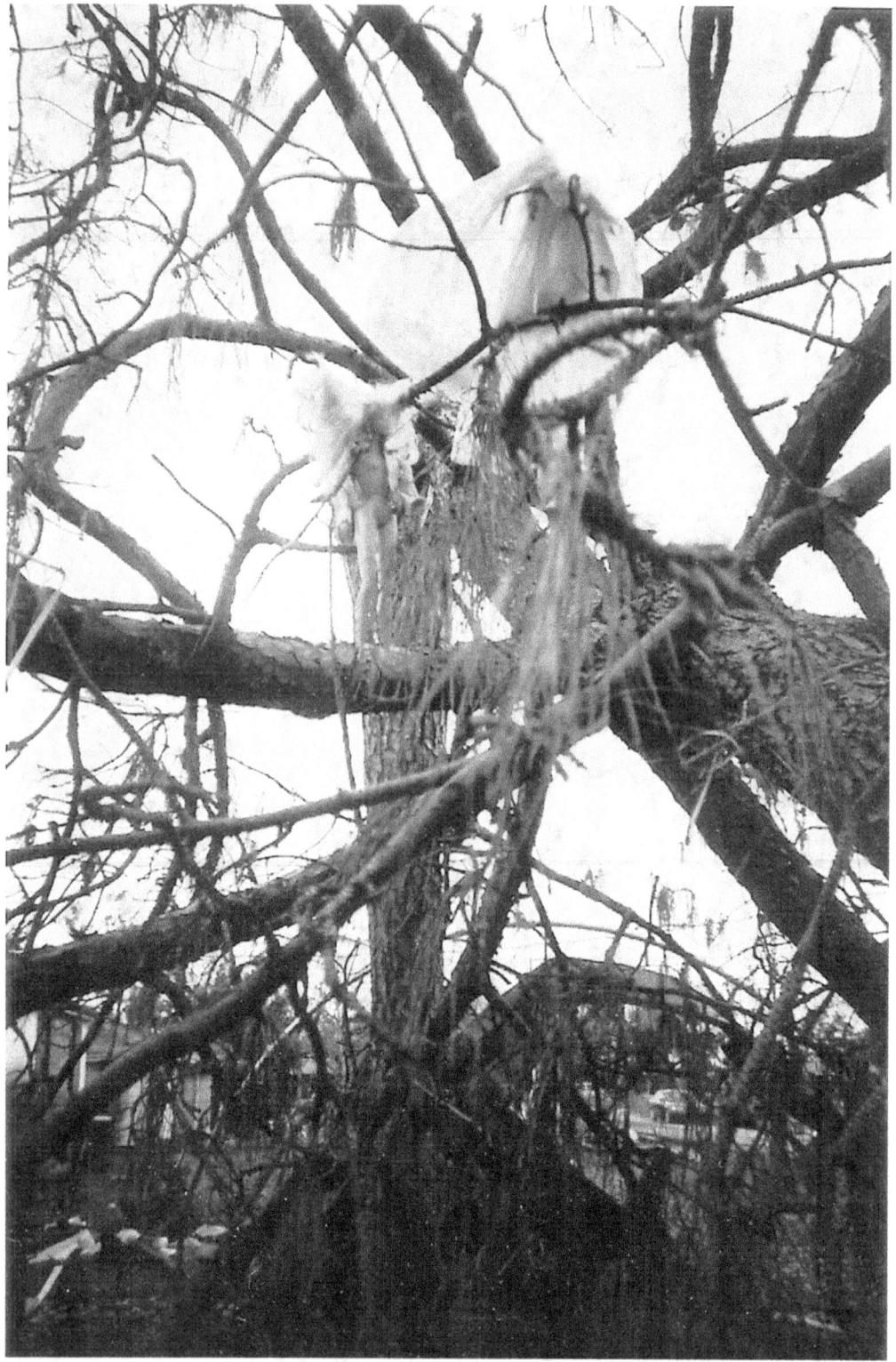

Barbie doll hanging in a tree

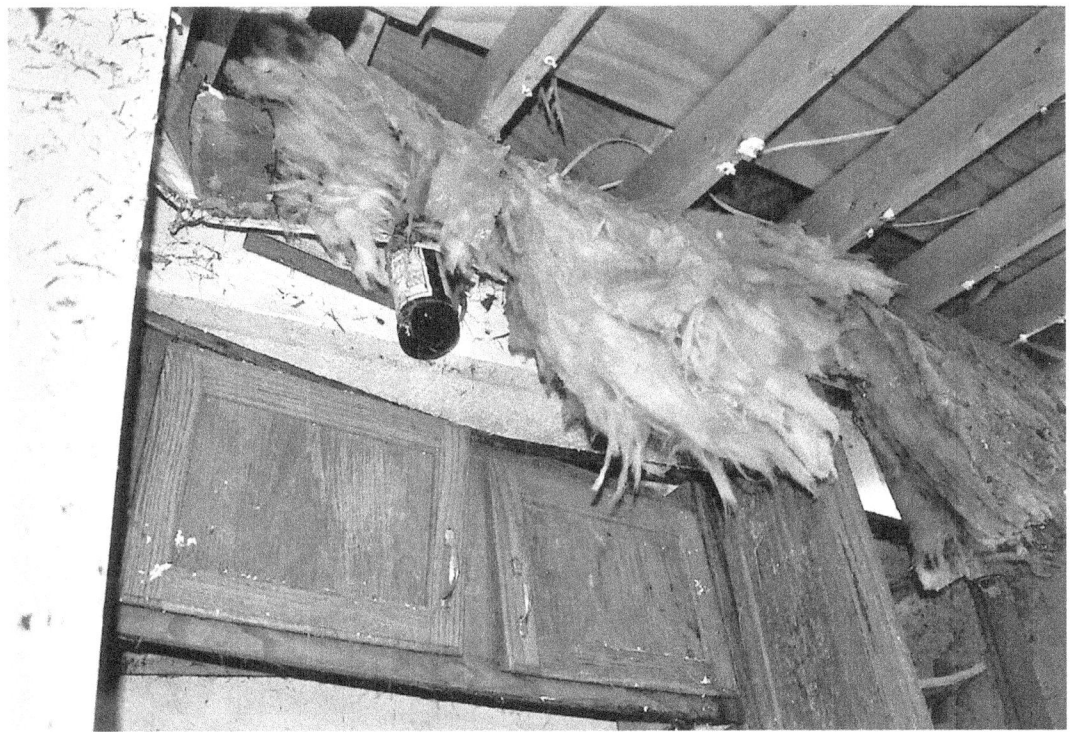

Bottle stuck in ceiling

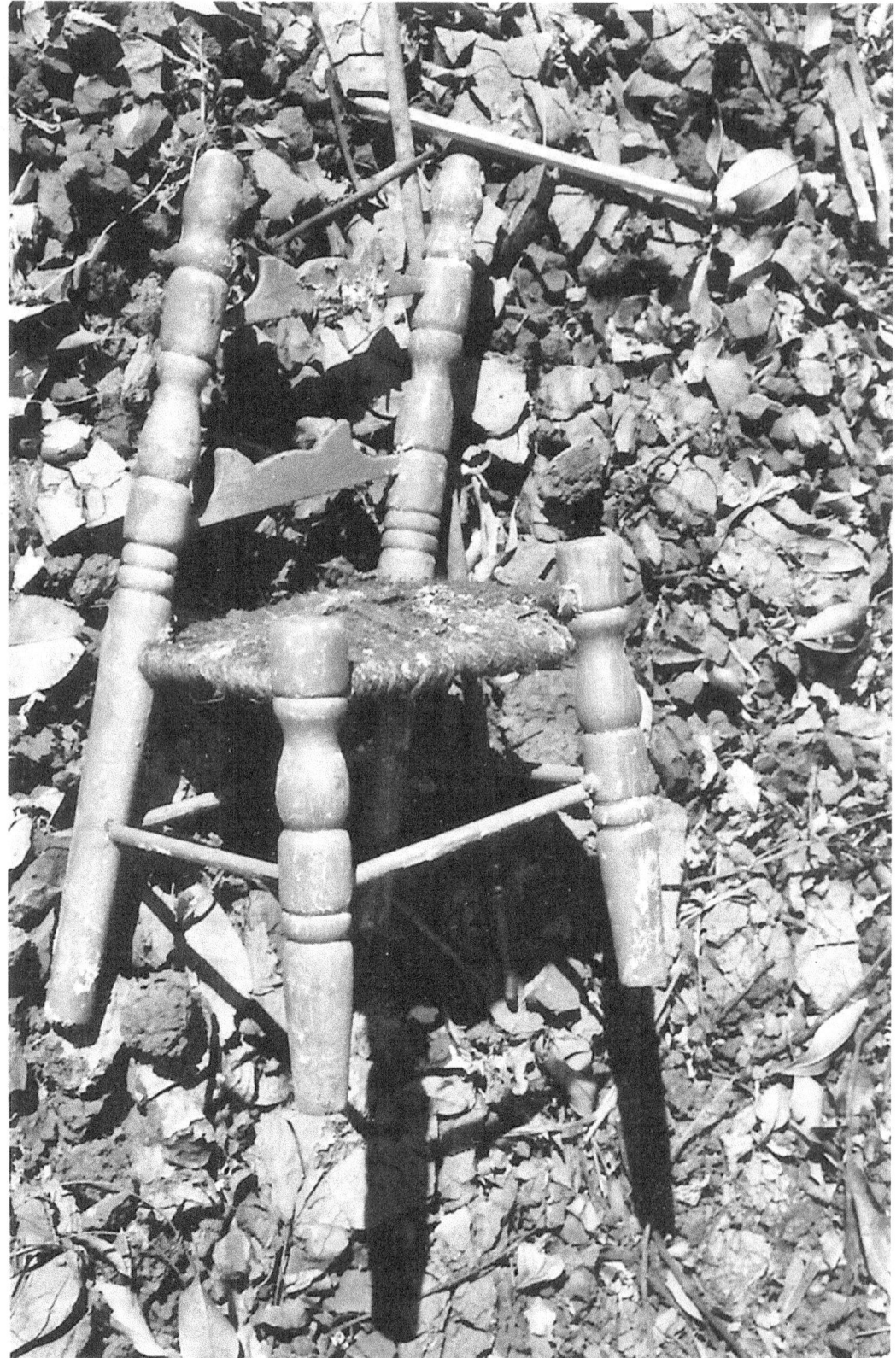

Child's chair on the mud

It was hard enough for people to leave everything that they had worked for, saved, collected and valued, but for the children it had to be worse. Adults could understand what happened and why, but children could not. All they knew was that they suddenly had nothing, no favorite toy, no clothes, no books, no games, no bedroom and no home. All they would remember was that they had to leave quickly never to return. What they would learn from this experience would depend on what they saw the adults doing. If the adults that they depended on were angry, they would learn anger. If the adults were lost and confused, they would learn fear. If the adults joined together with their families and planed for a new future, the children would learn acceptance. If adults worked to help themselves and others, the children would learn compassion. It is not just our futures that will be at stake, but what we do now, and how we do it will affect our children and future generations. The impact will be long lasting and worldwide.

Before we left, Nicolle wanted to show me some of her neighbor's houses. One across the street had left the front door open and the destruction inside was the same. We could hear an eerie squeak, squeak as a slight breeze blew the screen door open and shut. It sounded like the ghost towns in the old western movies. This was a ghost town, brown, silent, empty. Near the door was a photo album, tossed aside as if the owner was going to take it, but changed his mind and threw it down in disgust. In the yard on a tree that was downed, there hung a Barbie doll, staring, waiting for her owner to retrieve her. A gray stuffed elephant lay on the ground, its eyes focused on the sky watching to see if the storm was over.

When we left we decided to drive through Slidell to go to my Lacombe house before going to Baton Rouge. The bridges from New Orleans East to Slidell cross over Lake Pontchartrain. The twin spans on I-10, are similar to the Causeway, but shorter, and there is the old bridge along side the railroad tracks that is the Highway 11 bridge. The twin spans were damaged, unusable and closed, the storm having knocked out sections of it just like Hurricane Ivan did to the Pensacola Bridge the year before. But the old Highway 11 bridge was deemed to be safe. I thought it strange that all of the "old" structures did not suffer as much damage as the modern ones. Especially the train tracks. Perhaps structures were built better "back then." Perhaps newer materials were not as reliable as they were supposed to be. Or perhaps, as we were discovering with the levees, structures were not built the way they were supposed to be built, with companies "cheating" in places that they thought no one would notice. Then, an unexpected event uncovered the "mistakes" that were made, and we were left to suffer the consequences. Then everybody wanted to blame everyone else and we are told, "Sorry, but there is nothing we can do about it. It wasn't our fault, it was someone else's fault." The buck never stops anywhere, but travels around and around and around.

Crossing Highway 11 took over an hour because of the traffic. On the Slidell side where there used to be fishing camps and summer places, there was nothing left but piles of broken lumber along the road. Where some people had come back to clean up, their entire camp, furniture, belongings, inside and outside, was nothing but a giant pile of rubble on the side of the road. In the water were stumps where the camps used to be. On the road and even on top of trees and broken houses there were boats that had floated up with the water. Right in the middle of all of this destruction, there was one camp that was completely intact. It looked as if it had just been built. There was no damage at all. The camp looked clean and new, the paint appeared to be fresh, no windows were broken, no apparent water or wind damage. The more areas that we went through, the more unbelievable it was. It was hard enough to understand that one area was impacted and devastated, but neighborhood after neighborhood, house after house, family after family, thousands and thousands of people and structures destroyed. Yet among all of that devastation, one house, one person, one thing seemed to have survived with no damage. How could that have happened? Could anyone answer that question? Even scientists cannot explain why one house, or one tree, was completely devastated and the one next to it was totally intact.

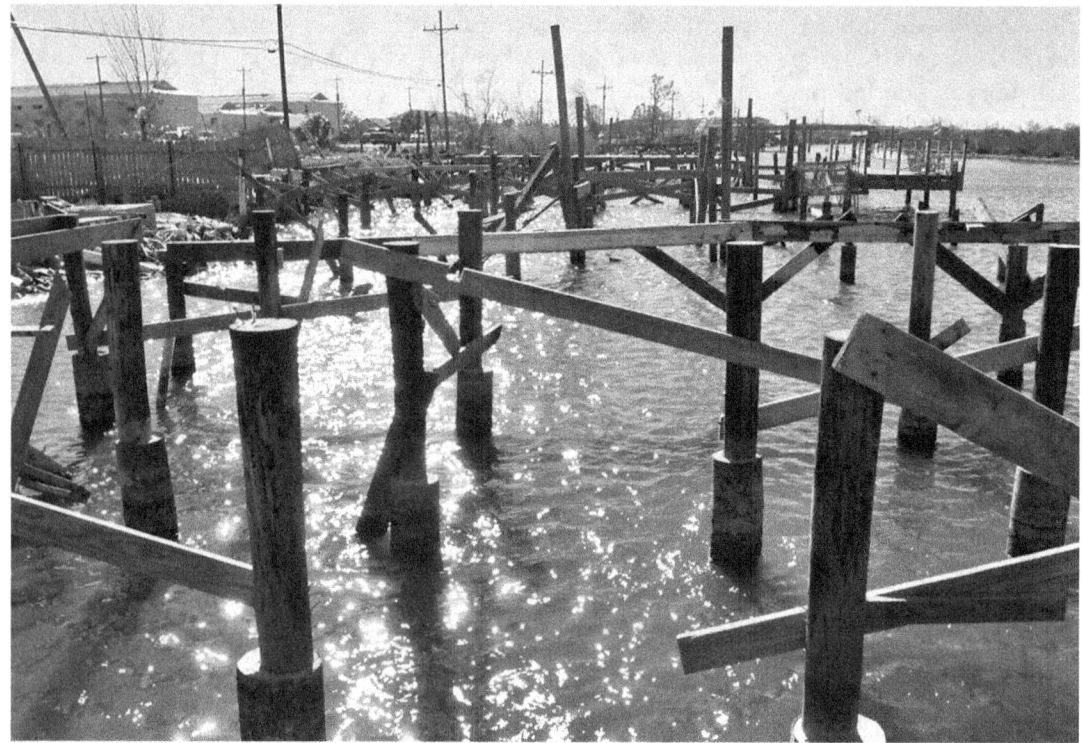

In Slidell where camps used to be

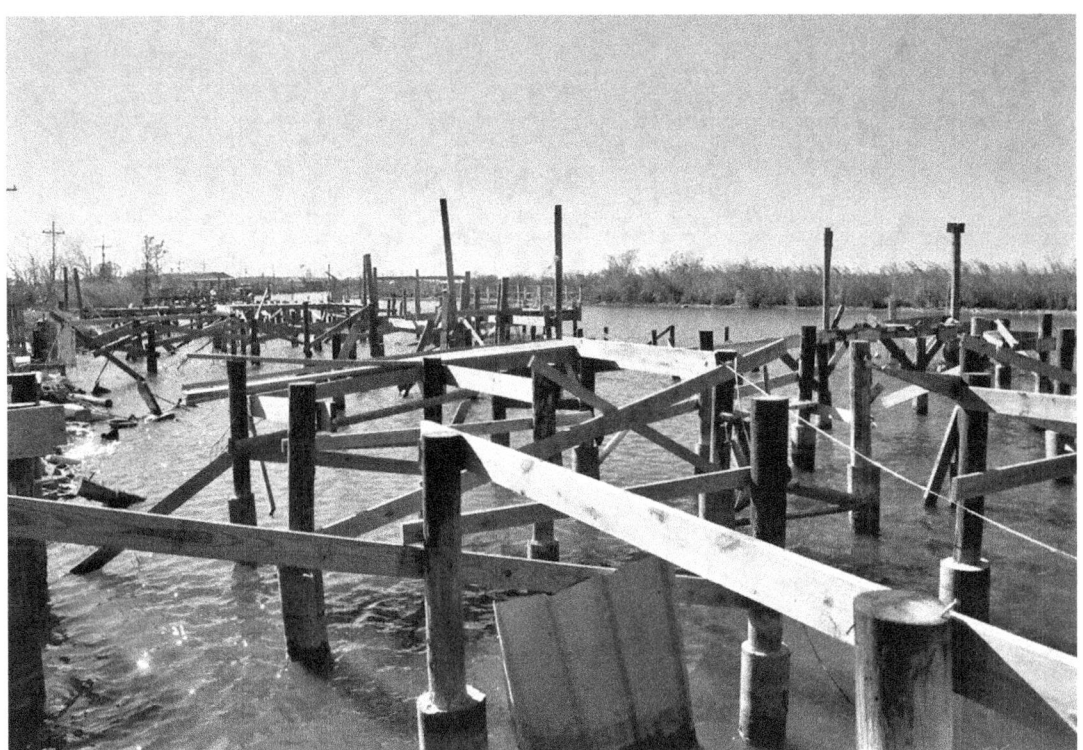

Empty pilings from camps

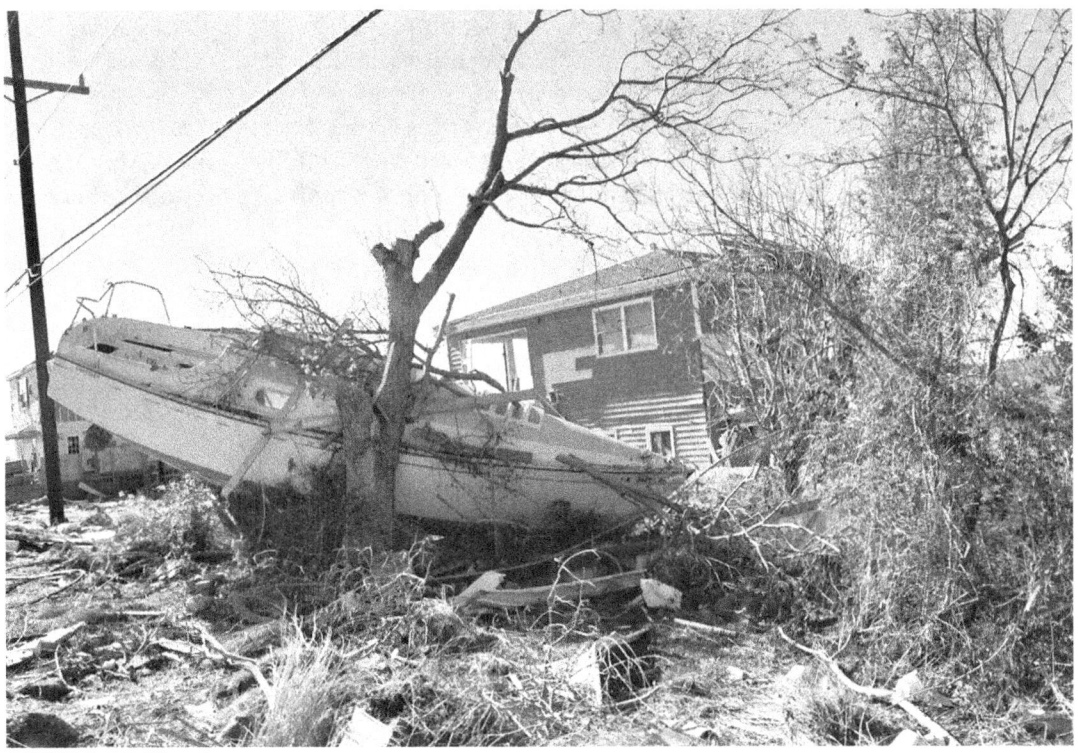

Boat landed near a tree

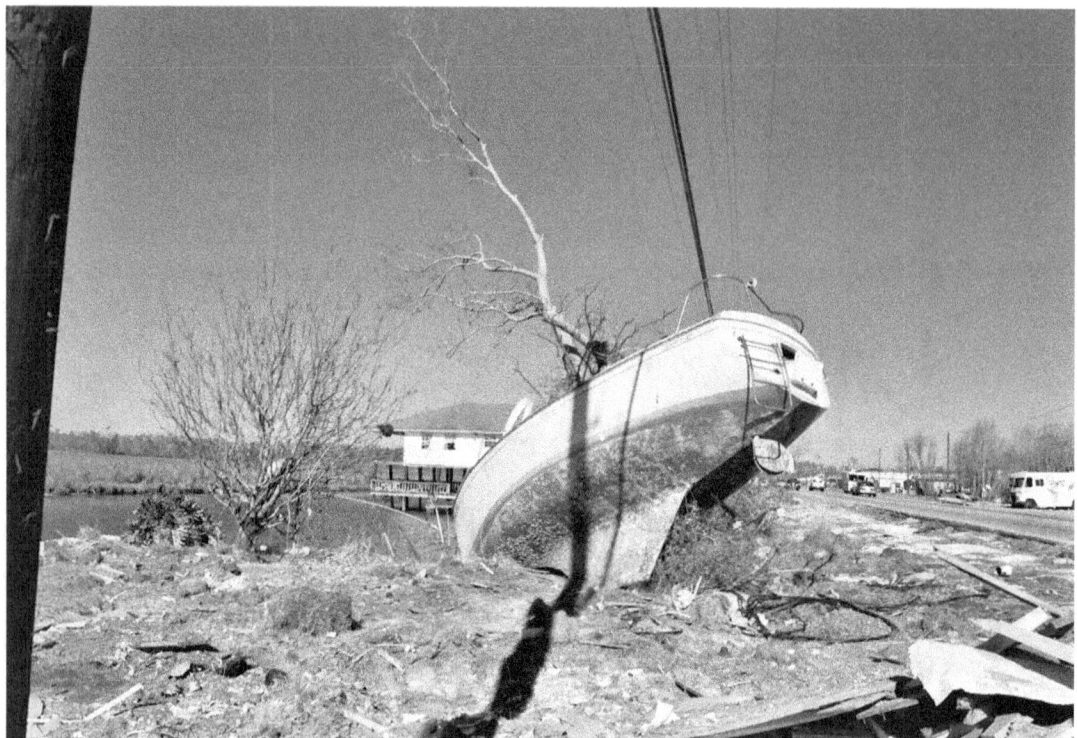

Boat in debris

How could I describe the scenario without repeating myself? Everywhere we went there was more of the same. More lives displaced, more houses destroyed, more devastation. Perhaps if I said it 500,000 times, that would be how many houses were devastated. Perhaps if I said it 1,000,000 times, that would be how many lives were affected, displaced, and lost. There are many more places that I had not been to and never will see. That is what people who heard about this cannot understand. There is no photo, there are no words, there are no feelings that can communicate the entirety of the conditions here. Snapshots and stories are only small parts of the totality of the ruins. Each story, each photo, each description, and each feeling has to be multiplied a million times in order for anyone to understand, and yet it would remain an unbelievable incident. People might be reminded with another story in the newspaper or on television, but soon it becomes a second page item in the storage system of the brain, replaced by the latest headline, or current happening. We, here, amid all of this mess, have to live with it every day. We see it all around us, we hear it in the machines working to clear it, we smell it still, we meet others and talk about it constantly. It will not go away. We try to forget for a few moments by going to work or shop, but there is no forgetting. We live it. It has become our new way of life. We have to deal with it. We have no choice. Even those who have decided to move away from it all, it will be forever in their thoughts. It has become a new memory, an unpleasant one.

I was not sleeping very well. I would wake up two or three times during the night, then would be too tired to get up when daylight came. Could it be the heat, with still no electricity to cool off with? Even with a generator, I conserved because of the amount and cost of gasoline to run it. Could it be that I stayed up late looking at the stars? There was not much else to do except eat, sleep and clean up. I spent the days eating at Rita's and watching episodes of "Firefly" on DVD, taking my mother places, occasionally riding to Baton Rouge to visit Nicolle and/or wash clothes, taking Cherie to dialysis when she stays with me, and meeting my writer friend for lunch or dinner at one of the few places open. There was no schedule to follow because there was no regularity. We waited for services to be restored so that we could resume some kind of plan, or stores to be open on a more regular basis so we could get supplies without having to check on their limited hours.

I sat on my balcony in peace and quiet. It was a clear day with a slight breeze coming from the north. I heard some traffic on the highway, and some hammering in the distance. Crickets were constantly chirping and birds were calling here and there. As I looked around my property, trees were down and branches were everywhere. If I closed my eyes and felt the sun on my skin I could pretend that nothing happened and life was beautiful. But my mind knew differently. Much of my family was displaced, confused and suffering. Some people said that God's wrath was sent to a sinful city, but we were not all sinners. Or everyone was. What person or religion had the right to judge who was a sinner and who was not, or what act was a sin and what was not. Other people said that it was Global Warming that has caused this terrible act and we were all to blame. The scientists don't all agree with that scenario either. Some believe that the earth is going through natural changes without any influence from us. Those who believed that we were to blame said that we should change our ways before it is too late. What would they have us do? Should we give up our luxuries and electric toys that give us our freedom and pleasures? Big businesses had encouraged us to have these things. Big businesses would be the loser if we stopped buying luxuries. They won't let that happen. They tell us that they will improve our toys and cars to be more energy efficient, and encourage us to buy more and better items. They hold the key, but they are directing us to the wrong door. We, the little people, the consumers, were left with a "damned if we do and damned if we don't" choice. It was and would always be about money and power, and those of us who have neither, which is the vast majority of Americans, are left in the dust of this technological revolution wondering how it all got to be this way. And yet, we were to blame, so they say.

Where do we go from here? No one has the answers. Everyone has the answers. Religions say to repent. Politicians say rebuild. Scientists say beware. We are lost and confused, drowning in a sea of mixed emotions and ideas pulling us down from all sides. Nature would renew Herself, as Nature always does. What would Man do?

5.
Renewal

Blue roof.
Bigger and better.
Clean up.
Closed.
Debris.
Devastation.
FEMA.
Government.
Gut out.
Help.
Homes.
Insurance.
Lines.
Lives.
Mold.
Open.
Politicians.
Raze.
Rebuild.
Red Cross.
Reform.
Return.
Temporary.
Trailers.
Wait.

This was the new vocabulary of the South. Every sentence spoken by us contained one or more of these words. Every person who spoke them did so with either anger or hope, sometimes both. The small areas that were progressing toward normalcy were the ones where the individuals were angry, but they were also demonstrating hope by doing what they had to do to reclaim their lives and property. Where the government and politicians were working, there was nothing happening but talk. The people in neighborhoods that had something of a house left were out there with their shovels and gloves and were getting rid of their mess on their own. They were saving what they could with the help of their families and friends. The longer they would have waited for help from government agencies, the harder it would be to clean up as the mold would continue to grow until nothing was salvageable. The mountains of debris in front of their homes reflected the fact that someone was there working for a future, they were mountains of hope.

Mountains of tree debris kept rising in Lafreniere Park and other large areas that used to be green spaces for people to enjoy and relax. These places were closed to the public indefinitely. I tried to go into Lafreniere Park to take pictures of the mountains, but was chased away by the police. There was not even that small pleasure left for me in Metairie, only the memories. I remembered the dream of Ricky telling me not to go back to Metairie. Was he telling me that the past is over and not to spend time dwelling over what was done, but to move on toward a new future? Maybe that was the message for all of us, not just here in the devastated area, but all over. The only way that kind of message could reach out to the world was through an event so unbelievably huge, that it was impossible to ignore. Our evacuees were dispersed all over the country, each one crying out for something. A million people, each affecting a million other people who were trying to help, each in turn affecting a million other people who listened, cared and offered something to ease our pain. All kept asking how could this have happened, and the government was investigating and studying the issue to find out how and why. But the past is over. Sure, we could use that information to improve the future, but the bigger question we should all have been asking is what to do next. Let the lawyers and the politicians play the blame game because they were the only ones that would benefit from it. None of the people that they represent will end up with anything worthwhile. Lawyers and politicians could sit around for years and years fighting this because they had the power and the resources, but those of us who needed the help would not see it soon enough to make a difference.

Something had to change. A lot of things must change. This country, this world was headed in the wrong direction. The rich, the powerful and the greedy were the only ones who were getting richer, more powerful and more greedy. And they were getting there at our expense. They lied, cheated and stole from us and made laws that made it legal to do so. They force-fed us their beliefs and punished and ridiculed us if we protested. We could just barely get by on our little incomes to have a little of our dreams, just enough to be reasonably content, but not enough to feel totally fulfilled. Something was wrong with a system that let us reach out for a better life, but stopped us just short of our goals. It was like the carrot dangling in front of our noses, and the harder we tried to get it, the further away it got, until we died trying.

I couldn't keep track of the days. The days and weeks had been going by so fast that I had to look at my cell phone for the date. Even the months were flying by. September was over already and it was October. Maybe it was because we kept anticipating good news, and when it would come in little trickles it kept us hoping that more would come soon. If I looked back at what we had accomplished, there did seem to be a huge improvement in our lives. In October, Nicolle went to her home in St. Bernard Parish a number of times to clean out the mess and find more to save. She had kept some of the boys old toys and school papers in plastics bins in the attic, and the water didn't get inside. The more she looked, the more she found, and the more she found, the more she wanted to look. Her husband, Charles, had an antique car, a 1970 Road Runner, that was his prize possession. He took it to many a car show and got numerous awards for it. It was left in their garage and spent weeks under water, then weeks before they could get an insurance adjuster to access the damage. In the meantime, he couldn't touch it. After they came out to see the car, Charles could move it someplace where he could try to clean and restore it. The perfect place was on my property, since I have so many acres and a huge garage that he could keep it in. Then, since I had a place for them to keep things, they started to retrieve more things from their house. They could not have kept things at the house in Baton Rouge where they were staying, because it was already crowded with eight people, three dogs, and a cat or two. Nor did they know how long they would be able to stay there. Nicolle and her in-laws had applied for a FEMA trailer and both were approved, but they had no idea when they would get one.

With Charles' car here, they began to spend more time at my house. Since I finally got my electricity on, the boys and Nicolle could use my computer. They would spend weekends here enjoying the peace and quiet of the park-like setting. They would work and play amongst the tall pine trees in the spaciousness of my property. Even their little dog, Dox, liked to run around chasing the squirrels. It was like a vacation for them. It was good for them to get out of the cramped house for a while, even if just for a little while.

Charles had returned to work, but he had to go to the Kenner store near the Louis Armstrong Airport. Staying in Baton Rouge, he had to drive ninety miles to work and back each day. Because of the traffic, the drive took two hours each way, so he was putting in a twelve hour day. Like most people who got their jobs back, all they could do was drive, work, eat and sleep. Sometimes since driving and working were a priority, eating and sleeping often got left out. When Nicolle's and her in-laws' trailers came, they were happy to be able to have "their own place" again. They chose to put the trailers near Picayune, Mississippi where Charles had family and lots of land. Nicolle was dissatisfied with the schools in Baton Rouge. The ones in Mississippi were near the Stennis Space Center where little Charles and Nicolas would be put in the kinds of classes they were better suited for. The schools seemed more sensitive to the problems the children faced since that area was hit harder by Katrina than the Baton Rouge area.

After FEMA checked and rechecked the trailers, they were able to move in. But what a surprise! They looked nice from the outside, but they shouldn't have called them trailers, when in reality they were campers. Maybe one or two people could fit nicely in them, but Nicolle's family of four was more cramped than in Baton Rouge. "Little" Charles was fifteen and taller than Nicolle, and Nicolas, who was twelve was almost as tall as his mother. "It's like living in a submarine," she said. So there was still no room to keep things that she salvaged from her St. Bernard Parish house. Well, I guess that's progress.

Charles still had to drive about the same distance to go to work, but in the other direction. He still spent twelve hours gone every work day until the twin spans from Slidell to New Orleans East were opened. Crews worked quickly to restore these bridges because they were part of the I-10, and traffic was reverted through I-12 causing major problems, especially since there were many more eighteen-wheelers delivering rebuilding products to this area. There were thousands of trailers and campers being delivered, hundreds of military convoys, convoys of power companies and tree cutting trucks and thousands of new workers who came here to help rebuild. So when both of the bridges of the twin span opened, the traffic congestion lightened up and Charles' drive to work was cut in half. But he still spent ten hours away from home each day.

Rita's life got somewhat easier, too. She found a new job, but had to drive across the Causeway, then across the river, also spending over an hour each way, just for a part time job. The money barely paid for her living expenses. Since all of her services were restored, and she got a new stove, she could do more cooking which she loved to do. We pooled our resources for food and I would eat over there since I didn't like to cook, especially for just myself. Mama cooked too and spent her time half at Rita's house and half moving back with me. They both helped me with the clean up process here and in my Metairie house, since I still had things there that I hadn't moved yet.

Alyssa had to drive over an hour to go to LSU in Baton Rouge for her classes since UNO was closed for repairs. She didn't want to lose a semester and found a few classes to help her towards graduation. But she then dropped a bombshell on us. Her boyfriend was in Japan in a program that taught English to Japanese children, the same type of program that Lauren was doing in France, teaching English to French children. Alyssa didn't want to be away from him any longer, so they decided to get married and she would go to Japan in January. Being the independent one, she worked out all of the details and got her plane ticket before she even told us of her plans. She would be able to finish school in Japan. Now with her two daughters gone, Rita had new decisions to make. She and Alyssa wanted to go to France to be together with Lauren and Kaylee for Christmas. Kaylee, who was eight, had picked up the French language so well that she spoke it like a native. They liked it there so much that Lauren wanted to stay in France. Our family had relatives there that they could visit. I thought the events of Katrina and the aftermath had been influential in their decisions to be away from here. The French medical and education systems were also factors in Lauren's decision to stay in France. Rita's daughters would be on opposite ends of the earth and because of the expense, she could not visit them very often. The same was true for me, since I was retired and had more time to travel. But for both of us, the expense of travel added to the decrease in income would mean that we would have to give up many things in order to afford to travel. It was a hard decision to make, in order to have the time to do what we really wanted with our lives, we would have to give up the jobs that paid us well enough. But if we kept the jobs for the money, then we wouldn't have the time to devote to our dreams of traveling, writing and painting. Katrina made us re-think our priorities. We could have lost everything like Nicolle's family did, or even worse, we could have lost our lives if we had made the wrong decisions. We all survived, but we all lost something, some more than others. Rita and I became more determined to do the things that we had dreamed of all our lives, but never had time to do them. Because we are older, (I am ten years older than Rita,) there was more pressure to fulfill our dreams soon, or spend the rest of our lives working but never accomplishing what we felt was our true reason for living. We had just experienced the worst, the possibility that it all could have ended, then we had to struggle to keep what we had, living for a while with no electricity, no ability to find out the fate of our families and friends, spending hours in lines for food and gas, spending hours on the road for our bare necessities, waiting days and weeks for some kind of help, and getting little, doing what we had to do for ourselves. Our lives would be different. Katrina had given us a new urgency to live better and make every moment count.

I had already begun to change my life. I had already decided to retire to this new place, my peaceful and quiet retreat to do my art work and write. I had already given up a comfortable but small salary in order to devote more time to myself. After Katrina, I knew what it was like to be totally confined and alone. I knew the fear of the "what ifs." What if I can't get out to get Cherie to dialysis, what if I run out of food, what if I can't find my family, what if everything was taken away from me. But instead of wallowing in the "what if" mode, I had to do something. I could not afford to sit around blaming others and getting nothing done.

The thing to do was to establish a routine. But that was difficult because as soon as I had a routine, it changed. I spent weeks in a routine of starting the generator up first thing in the morning so I could have coffee, then turning it off to save gas, then cleaning and arranging things in my house, going to Rita's for dinner, writing down the events so that I would not forget them, then starting the generator again for a little while in the evening for a little light. When my electricity was restored, I had to develop a new routine. I could spend more time at home, especially since I could use my computer in the comfort of air-conditioning. I began transforming my notes to this story. I could contact friends and family by e-mail (my land line still worked only intermittently.) The insurance adjuster finally showed up, but there was a long wait for the money. The check was "In the mail" for weeks. Then I could replace the washer, dryer, refrigerator and stove. I hooked them up in the garage to use since I had not fixed up the kitchen yet. Nicolle would come over once or twice a week to wash clothes for the four of them. She was spending $12.00 each time she went to the Laundromat. That added up to over $50.00 a month that she couldn't afford. Each little improvement in one of our lives would also be a big help in all of our lives because we did the best we could to share and help each other. Rita and Mama would cook for all of us and Nicolle would take home what was left. No food was ever thrown away. Cherie spent most of her time at her trailer because she did not have damage and got her electricity restored sooner than the rest of us. But her dialysis clinic in Metairie was shut down indefinitely. She continued to go to the clinic in Hammond but that was forty miles each way, three times a week. Even if she stayed with me, Hammond was forty miles away. With the gas prices still rising, she was spending a lot of time and money just to get to her treatments. Because she didn't have much damage, she didn't qualify for any kind of extra aid. In fact, after she received the extra food stamp disaster assistance, they decreased her food stamp allowance to $30.00 a month. Even with the little she does eat, $30.00 a month is a pathetically small amount. Being further away from me, I could not help her as much and the long drive and extra expenses were taking a toll on her health.

My Aunt Audrey and Uncle Marion, whose house was flooded by the 17th Street canal levee break, went to live in Texas with one of their daughters. They haven't decided yet what to do next. They are both in their 80's and will probably need some kind of assisted living arrangements soon. One of their daughters lived in St. Bernard Parish near Nicolle, and lost everything. That husband's family, being from St. Bernard Parish, lost thirteen homes. Many of the people in St. Bernard Parish who lived there for more than forty years went through Hurricane Betsy in 1965 when many homes were flooded because of a levee break. For some, this was the second time they had to experience total devastation. This was true for Nicolle's in-laws. The people in that parish, as well as the other parishes south and east of New Orleans and along the Gulf Coast, were faced with the decision of going back or re-locating. No one had answers for these people. If they went back, there was a very strong possibility of another hurricane destroying their lives again. If they decided not go back, they would have to start a new life somewhere else. Then they would have to establish an entire new way of life. New friends, new neighbors, new doctors, new neighborhood stores, new churches, new routines. Some people would choose to move and establish a new life, but most hurricane victims wanted to live their lives in their familiar surroundings. They lived in their own comfort zone and felt somewhat secure that they would be there for the rest of their lives. They were waiting for someone to either reassure them that they could go back and be safe, or that they should move indefinitely to some other place. But no one could help them with that decision. There were so many mixed messages being sent out to the public that no one knew who or what to believe. The politicians kept telling people to come back and come back now. The engineers and scientists were saying it was not safe for the next or any foreseeable hurricane season. Environmentalists cannot agree on if there were, or were not toxic residues to be concerned about for then, or in the future. People were demanding answers where there were none. With that kind of uncertainty, it was any wonder why anyone would want to go back; I don't think I would. Even, north of the lake, I was skeptical about my safety. Water came in my house and it could happen again. If I were told that I could not return to my new home, I would be angry and confused. The decision to plan a new life somewhere else, or return to a potentially dangerous place, would be a difficult decision. But the politicians seemed to want to force people to hurry up and make that decision. In some cases, entire families would have to relocate together or go their separate ways. Many in my family were already scattered all over this country before the hurricane. Some of my family lived within a hundred miles of each other, but they, too, had relocated in other parts of the country.

My Aunt Lou, who lived in Metairie and had two feet of water in her house, was brought to Seattle where her son's family is. She was also in her 80's and was severly depressed from the trauma of seeing her home damaged, that she didn't care about anything. All of her belongings were lost because the mold destroyed what the water didn't. She stopped taking her medicine, and if someone hadn't checked on her, she would have died. She was beginning to show signs of forgetfulness, and her son had some serious decisions to make about what to do with her because she would not be able to live alone anymore.

There were many stories like this of the older, ill or infirmed people who would have a hard time deciding what to do next, if they still had the ability to make their own decisions. Many were depressed because they lost their entire life's possessions along with their homes. They were displaced and their routines have been interrupted. No longer could they wake up in their own bedrooms, in their own neighborhoods, see their own friends, go to their same groceries or churches, or visit their same doctors. No longer would they have family gatherings at familiar places, in their own homes or those of their families. Holidays would be spent in strange places, maybe with family, maybe not. It was hard enough for healthy young people to start over after being plucked up out of one existence and dropped into another. It takes time to adjust, but the older people have no time. In reality these older people will not live as long as they would have before the storm, and some have already passed away. But would these deaths be listed as victims of Katrina, or of some other reason. If Katrina hadn't happened, many would still be alive in the comfort of their home with their family, friends, and treasured memories. Would they get any help besides a FEMA trailer before it was too late? Probably not. Their deaths would be recorded as normal due to the results of ordinary old age or illness. The real cause of the further devastation of their lives was society. Not the society of ordinary people, but a society created by government and big businesses who had decided that some lives are expendable.

The death toll by the beginning of 2007, directly related to Katrina was just over one thousand for the entire Gulf South Region. I suspect that it was much higher than that and will continue to climb. One unofficial report was that twenty thousand body bags were used. Another unofficial report was that depression was so high that the suicide rate was much higher than normal, even with the decrease in population. There were many more unofficial reports and rumors circulating around about the dead and dying.

One day, Rita and I took a ride to Lakeview to see the house where we grew up. Lakeview was a subdivision of New Orleans along Lake Pontchartrain where middle class and upper middle class people lived. We grew up in a middle class house on the other side of a levee from City Park. The water had gotten in to Lakeview from the break in the 17th Street canal, and every house was flooded. We could see the water line up to the roof of the first floor of the houses. We drove by our "old" house to see the damage. We often drove by there to see what the new people had done to it. They were always remodeling it and adding to it until it was almost unrecognizable from when we lived there. But all that was left was a shell. We took pictures and drove by the back of the house through the alley. We got out of the car to look around the property. They had been remodeling again before the hurricane. The windows were out and were stacked up in a back bedroom that used to be my grandmother's.

We decided to take one of the windows. The smallest one was a bathroom window that had six frosted glass panes surrounded by an old wooden frame, the original green paint still showed under layers of newer paint. We put it in the trunk of the car and drove away. We then drove to the lakefront to see more of the damage in our old neighborhood. We took pictures of the Coast Guard Lighthouse that was severely leaning from the wind and water. We saw many sailboats and other boats that washed up on the shore near the Lakefront Marina and piled up on the road, on each other, everywhere. Having seen enough, we drove back across the lake.

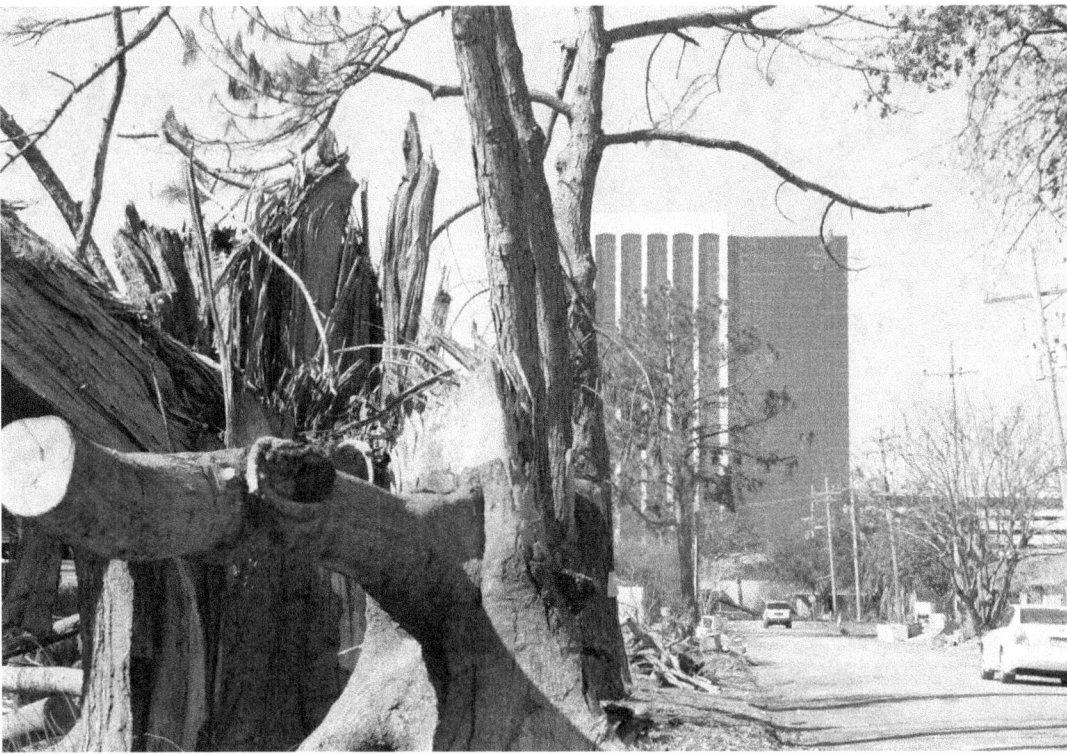

Tree damage near the 17th Street canal

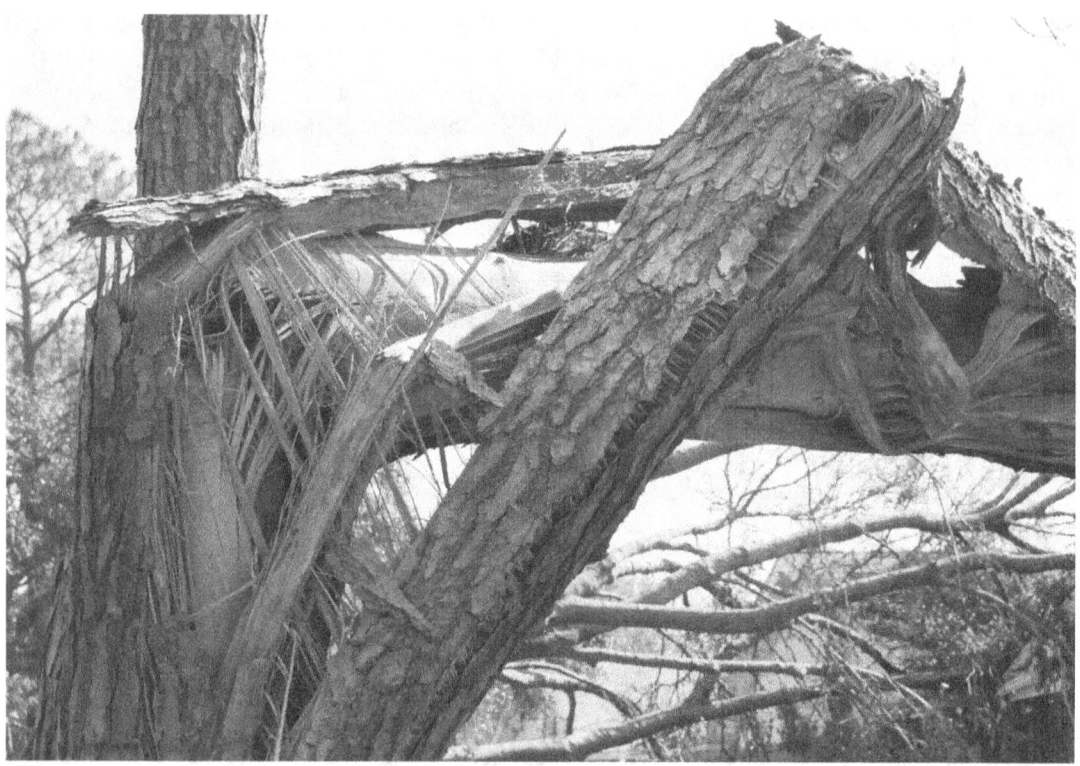

More damage near the 17th Street canal

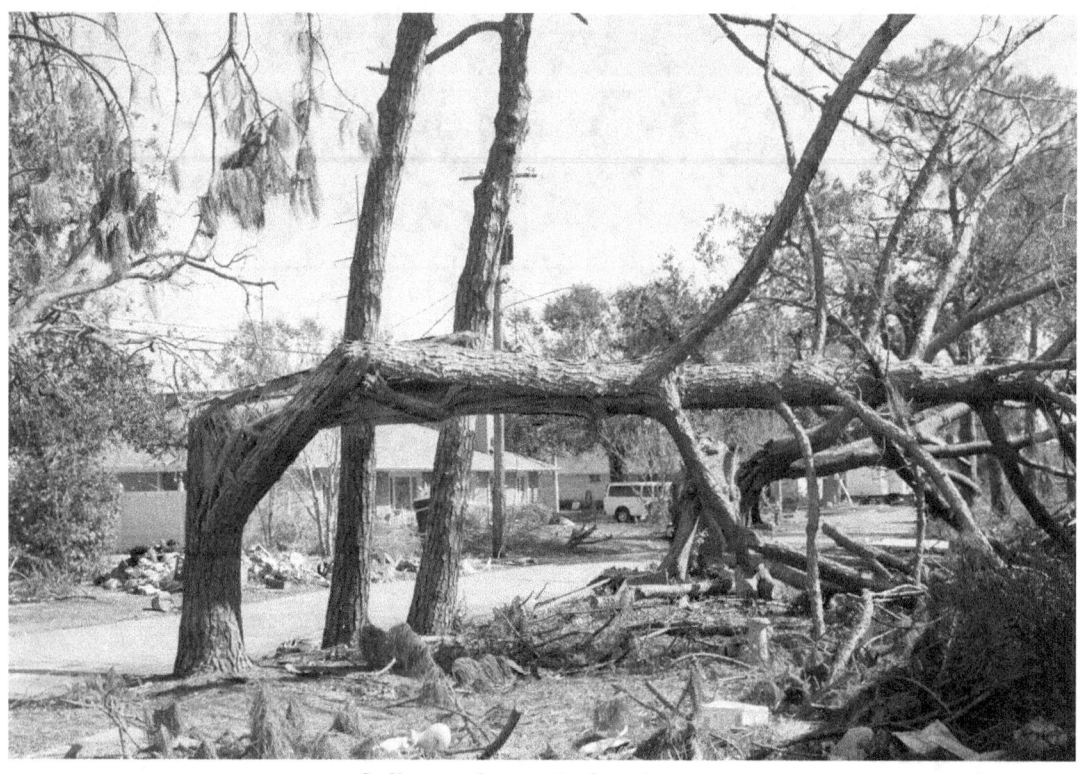

Split tree from wind and water

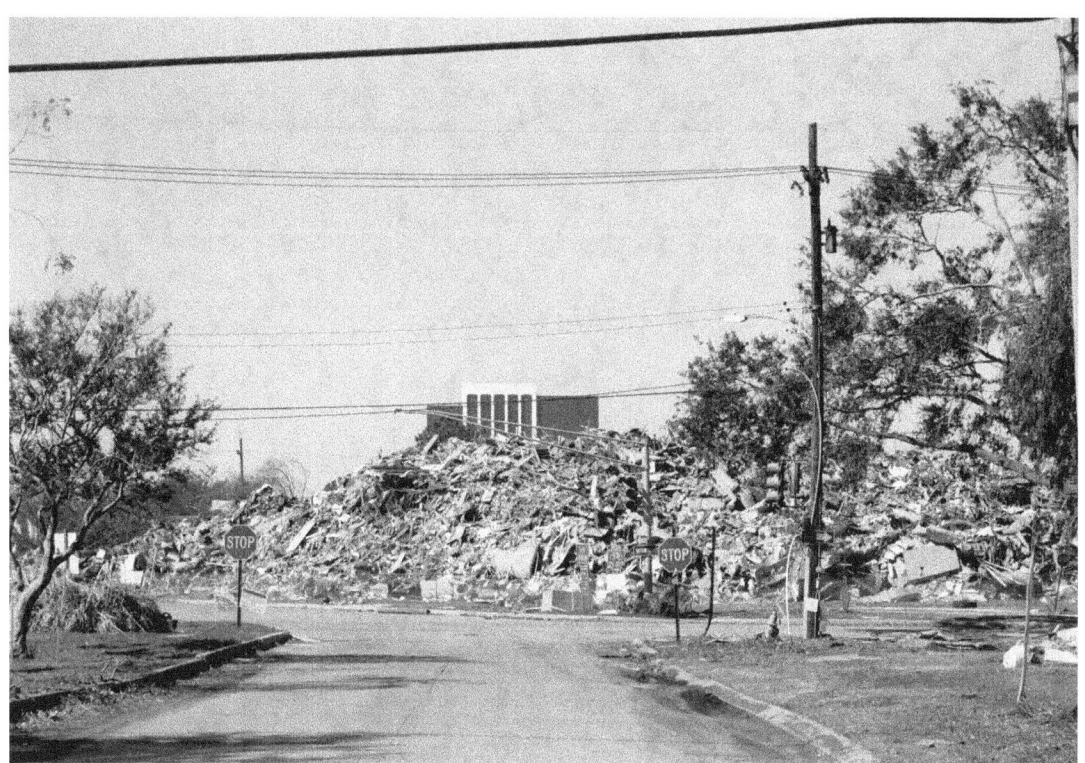

Another view of debris

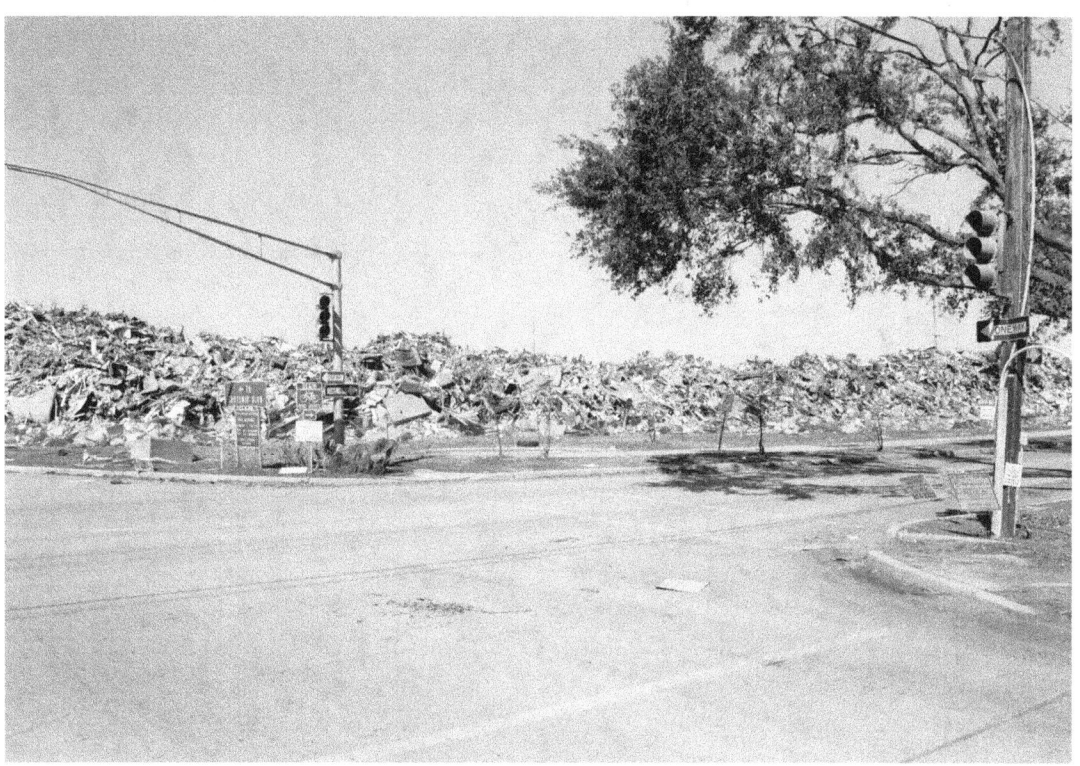

Debris piled up on neutral ground from Lakeview houses

Chandelier hanging from ceiling

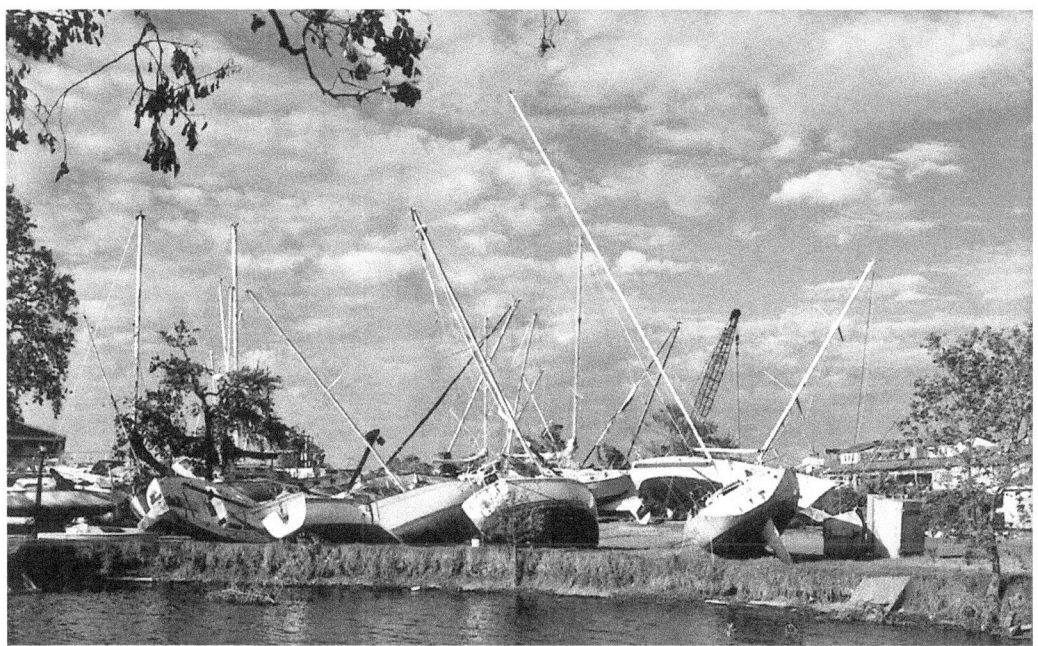

Boats washed up near the Marina

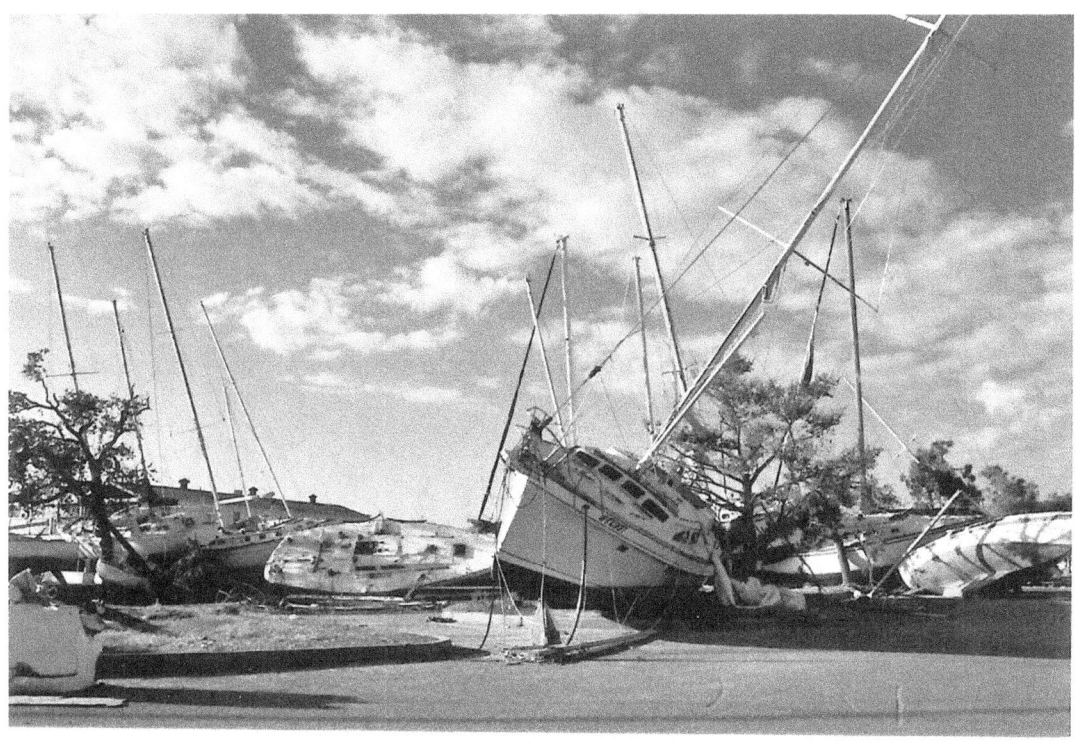

More washed up boats

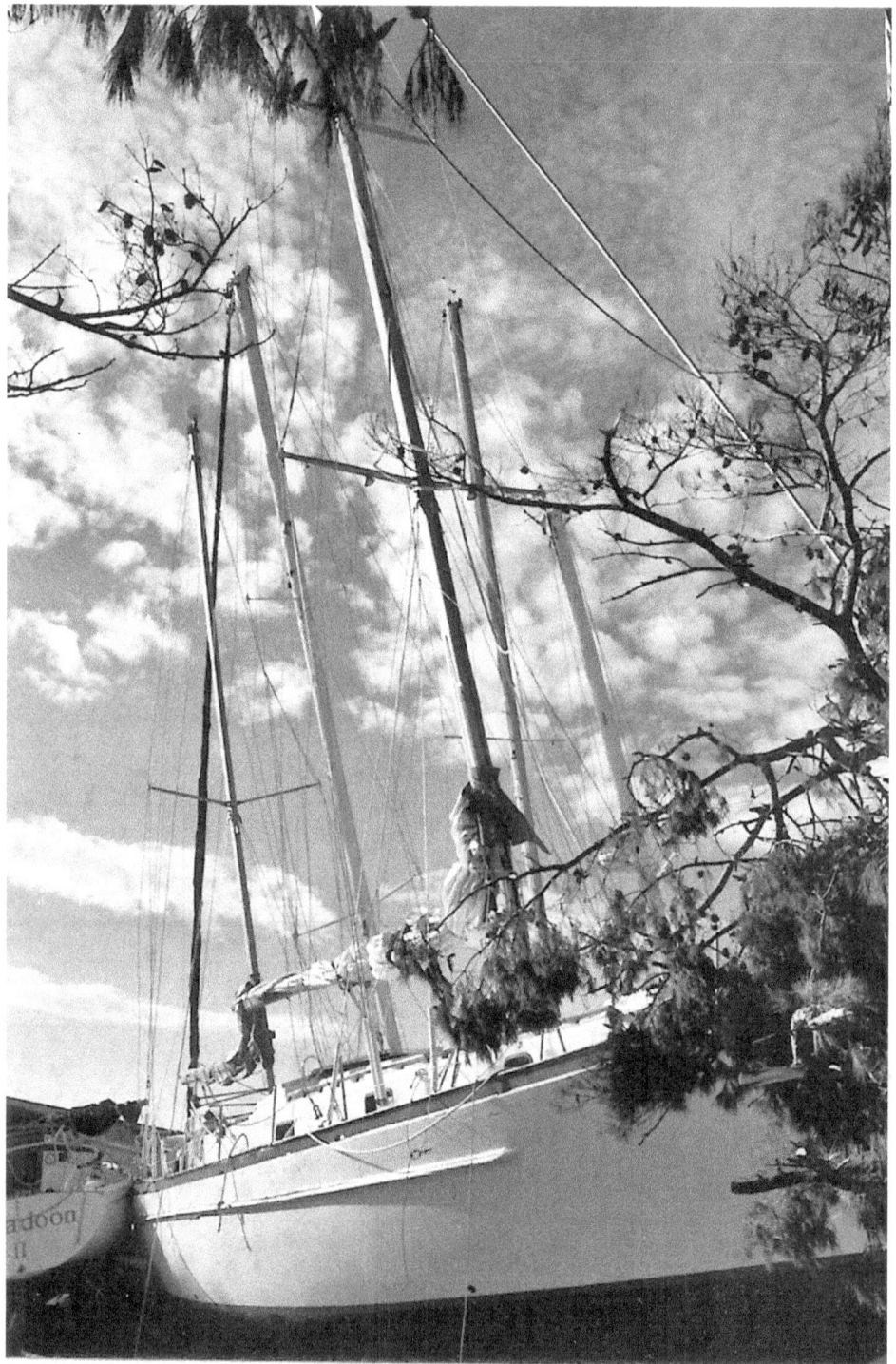

Stranded sailboat

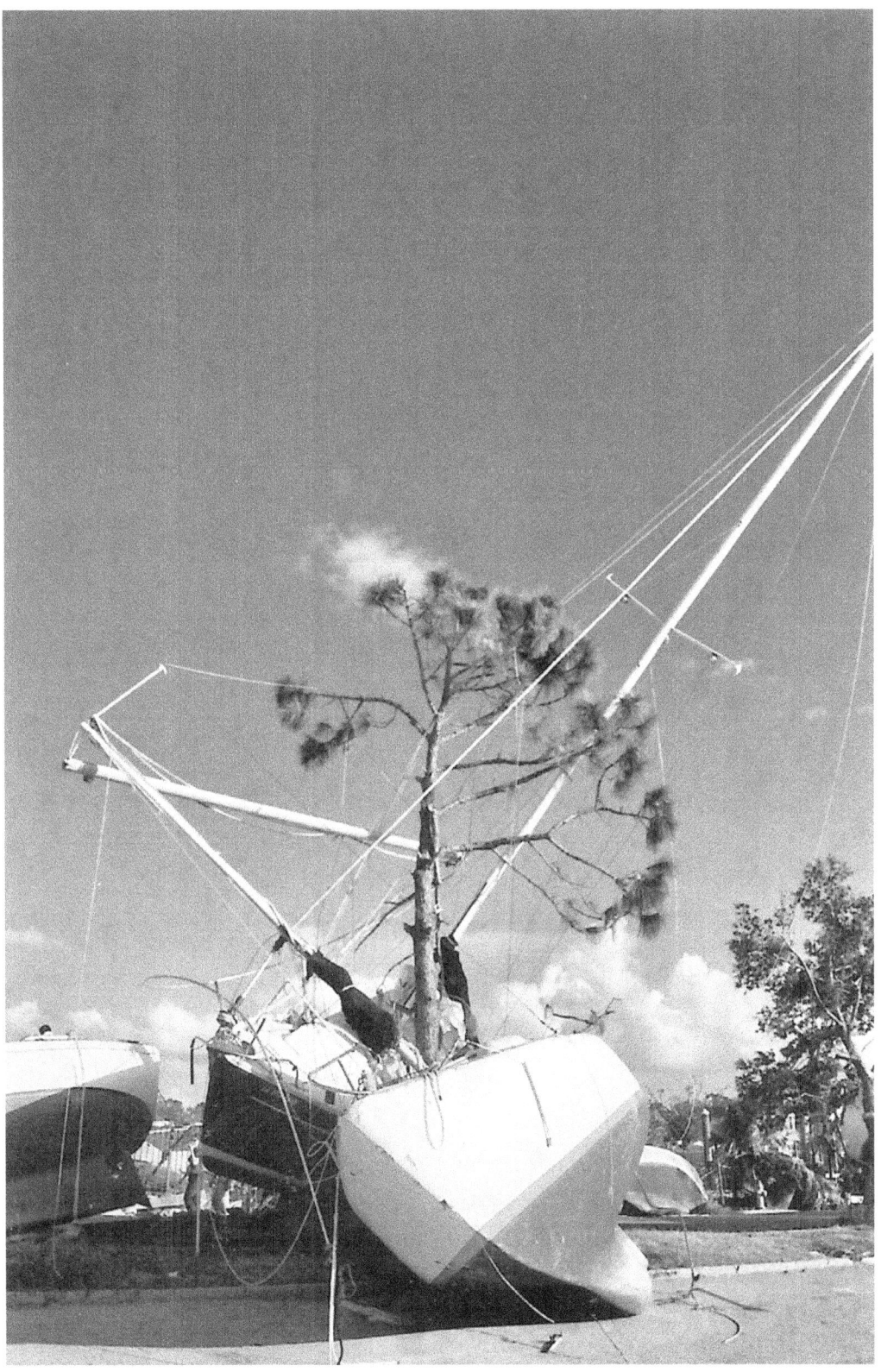

more stranded sailboats

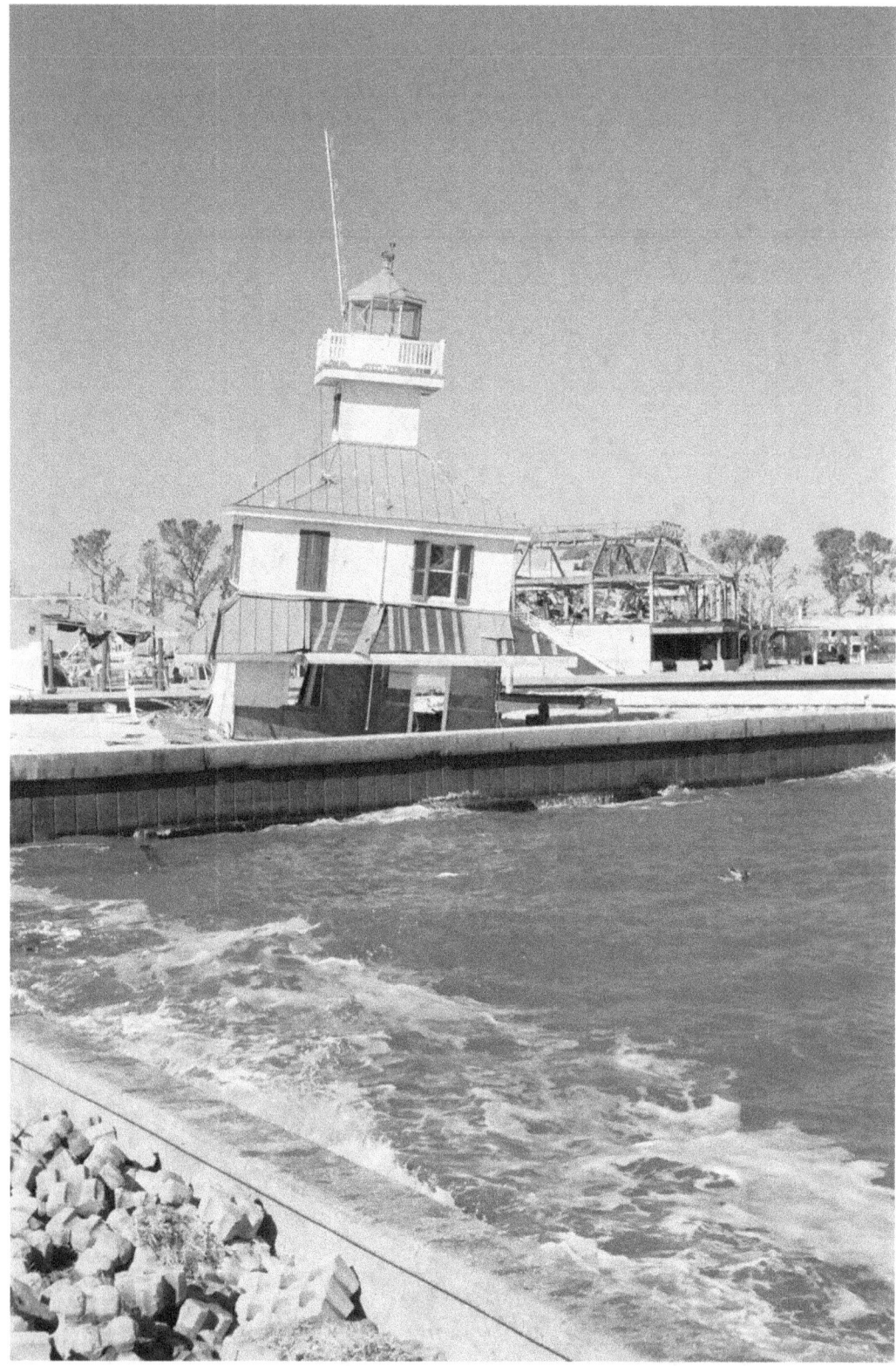

Coast Guard Lighthouse three months after Katrina

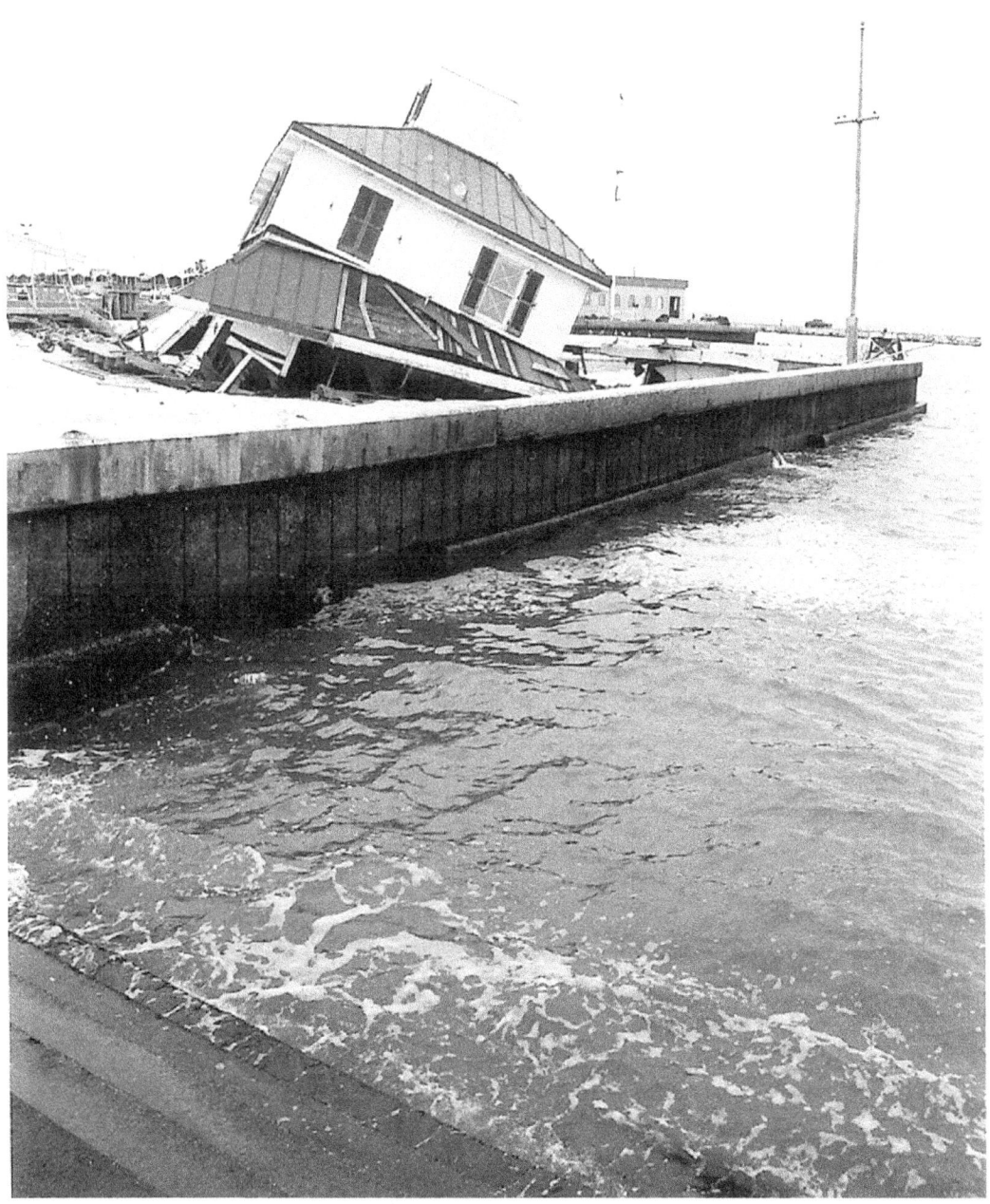

Coast Guard Lighthouse nine months after Katrina

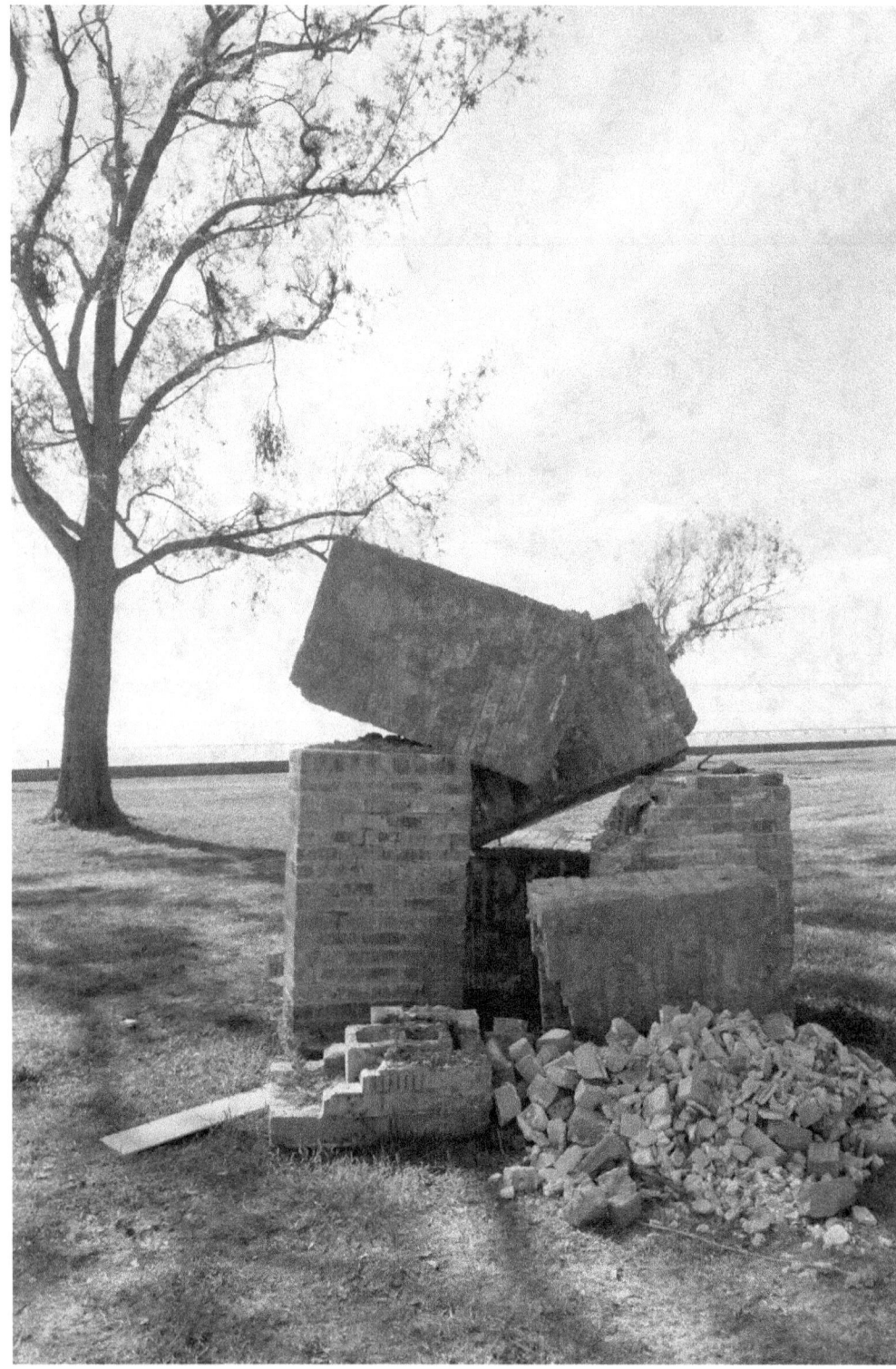

Brick BBQ pit toppled by wind and water in Mandeville

Often through the years, many of us took turns driving by the old house, each not knowing that the other did. Once, just before the hurricane, Nicolle also drove by there to see it. The remodeling had begun and there was debris in the back of the house. There were also some windows piled up in the trash. For some reason, Nicolle decided to take one of the windows then. What was it about windows that we all wanted to take one? Maybe one day, that reason will come to us. Perhaps we wanted a token to remind us of the happy times we spent there. Maybe it was a sign that we wanted to "see into" the past?

Growing up there, we all had special memories of the place. My children, Cherie, Ricky and Nicolle spent a lot of time there when I was taking education classes at UNO. I would leave them there so that my parents and grandmother could watch them until class was over. Cherie and Nicolle used to love my grandmother's room and spent many hours with Memere listening to her stories of the "old days," as she brushed their hair. Ricky loved following my father around. Pere, (my parents were Mere and Pere to their grandchildren) was a physicist and had lots of experimental equipment and machines in his garage. Ricky was fascinated by it all.

One of my children's special memories was when they spent time there in the summer while I was taking more classes at UNO. My father had a riding lawnmower to cut the grass on the three lots that were ours. He would hitch a little red wagon on the back of it and Cherie, Ricky and Nicolle would sit contentedly in the wagon while Daddy cut the grass. I suspect that after the grass was cut, he would continue to ride around the property because his grandchildren enjoyed it so much.

My father died in 1980 of pancreatic cancer. Since he worked with X-rays in the cancer center at Charity Hospital, we suspect that years of exposure cause his untimely death at the age of 62. Memere, my mother's mother, lived with us in that house since it was built in the late 1940's. She died at the age of 92, a couple of years after my father died. Then Mama was left alone. She made the decision to sell the house because there were many repairs to be done that she couldn't afford or handle. Before Daddy died, they had planned to move to Mandeville when he retired, but his death changed her plans. After Hurricane Betsy in 1965, Mama always worried during the hurricane season about the "big one" that was one day predicted to hit. Betsy, Camille and many other storms only added to everyone's reservations, even though these storms were never as bad as they were expected to be, at least not for us. Everyone silently worried each time there was a hurricane on the way, but all of the "near misses" added to a feeling of complacency. So in 1987, when Mama decided to sell the house, we had mixed feelings of sadness and relief. She lived for a time with Rita, and me, and Rita again until Rita bought a house in Mandeville. Five females lived there, Mama, Rita, Rita's two daughters Lauren and Alyssa, and Lauren's daughter Kaylee. Then I moved to Lacombe just before Hurricane Katrina, to live permanently in my "retirement" home. We all ended up on the north shore, but not the way we thought we would. Dreams and plans were always changing. We had to adapt to the changes that Nature and Man carried out on us. We constantly live with the "What Ifs" and did we do the right thing or not. When Mama sold the house we all wondered for years if it was the right thing to do, but when we saw the destruction of the house that we grew up in, we realized that in the long run, she made the right decision to move. What if she had not sold the house and stayed there during Katrina? The house and the neighborhood were devastated, and Mama would have lost everything. An uncertain decision in 1987, turned out to be the right one for her. We could keep our fond memories of that house every time we look into our windows from the past, instead of horrible ones of its devastation.

November. The holiday season. Not much has changed, yet everything is different. On Thanksgiving we were supposed to be thankful. We were thankful that we were alive and had a place to live. Ricky was born on Thanksgiving Day. He would not be here to celebrate his birthday. I always cooked for Thanksgiving, but that year I would not. Rita cooked a turkey and we went to her house to eat. There were only nine of us left. Nicole, big Charles, little Charles and Nicolas, Mama, Rita and me, Alyssa and Cherie. Lauren and Kaylee were in France. Other family and friends that used to come by to visit would not be with us either.

The day after Thanksgiving was usually the biggest shopping day of the year. There are sales everywhere that began at 5 AM. Rita convinced me to go early to get the computers that were on sale that I wanted for my grandsons. I never liked shopping, even for Christmas, but the computers were something special I wanted. I stood in line in the dark thinking I would go in, get the computers and go home. The people in the line were all talking about the effects of the hurricane and the problems they were having. That was not a very cheery situation for anyone. When we got in the store, they said that there were only five of the computers in stock, and they would not let us order one on line because it was an in-store special, besides the computers were down in this computer store and they could not order anything online. What progress! It only made me more annoyed that these big companies would do that to people, advertise a special sale just to get people into the store and not have enough for all of us that stood in line for one. I thought that the big companies did it deliberately, the old "bait and switch" routine. Instead of customers going away happy that they got what they wanted, the people would end up fighting for the last item. Did they do that to encourage fighting? They would get free advertising if it got in the news. Nothing like a good fight to lift up a person's Christmas spirit. I spent eight hours going from one store to another getting more aggravated each minute. Finally I went home and ordered the computers on line from another company, and got them for a better price. I should have done that in the first place.

Christmas. Another subdued holiday. Rita and Alyssa got cheap tickets to France to be with Lauren and Kaylee for Christmas. I brought them to the airport on Christmas Eve, my birthday. On the way back to Lacombe, I stopped by Rita's house to pick up her Christmas tree. She was not going to decorate, but decided to get a small, cheap "Charlie Brown" tree. Since she was leaving, I took her tree to my house because I didn't want to buy one either. I had a plastic white tree in the attic of my Metairie house, but when I unwrapped it, it was yellow. Either tree would exemplify the mood of the times, an old yellowed tree from the past, or a new but sad looking tiny misshaped real one. We celebrated Christmas dinner on the eve of Christmas Eve to be together before they left for France, but I also cooked a turkey on Christmas Day for the seven of us that were left. Rita insisted that I cook on Christmas Day so that we would not spend it as just another ordinary day. We exchanged a few presents, mostly necessities, not expensive superfluous items that are the usual fare, except for the two laptop computers I got for my grandsons. They lost their home computer along with everything else in the flood. Since Charles and Nicolas were in school, computers were hardly considered extravagant items anymore. Whether they would play games or do homework on them, they now had something new to occupy their time with.

2006. What would the New Year bring? A new beginning? New hope? New relationships? Changes? Definitely! Would those changes be for the betterment of the South and for Mankind? Or would there be more promises that would not be kept, like New Year's resolutions. I stopped making resolutions because I hardly ever kept them. It is like going to confession. "I'm sorry I did this and I'll try to do better." But nothing changes. Then, changes were forced upon us. We had to do things differently because our old ways have been taken away from us. Most of us had to change our entire way of life. We had to conserve because prices were going up, and we couldn't afford to live like we did before. The rises in prices blamed on Katrina were the gas prices, heating and oil prices and insurance costs, (if we qualify anymore). Who else would be conserving besides the poor and middle-class? Would the government be conserving? Would they stop the excessive spending in order to have funds to help out the needy? Many of us lost jobs and income, but I had not heard of any politicians, sports player, or otherwise "important" people losing their jobs. If they got displaced, they quickly had benefactors who helped them. Did the teachers, doctors, and ordinary, but more important, people get help to relocate? How many cuts in income did the politicians and sports persons get? I suspect none! How quickly did the rules change for the rich and powerful in order to get accommodations after the hurricane. How slowly, and how we had to obey the rules and get the red tape thrown at us before we got any help. How hard did we have to continue to fight for every little thing that we needed and deserved and even paid for? We spent hours even days on the phone trying get in touch with a human person to help us with our problems. Insurance companies made it the most difficult. With all their wonderful ads about how hard they were working to help us restore our lives, when we got a live person on the phone they curtly told us that they were sorry, but they couldn't process our information without this and that and when we got it to them, they need more. When we gave them more, they gave us less. Many wondered if this was done on purpose so that we would give up. Even the so-called "tax breaks" were so complicated and hard to qualify for I wondered if anyone was able to get any breaks. The policies were so complicated that I doubt even lawyers could understand them. These new rules didn't help the poor and middle class, especially those who paid little or no taxes anyway.

We were getting the message that some people were deemed more important than educators, doctors, scientists, and blue collar workers. Educators touch the lives of everyone, but educators were the lowest paid of professional people. That was true even before the hurricane, but only got exemplified because of it, especially in the south. (Because I am a retired teacher, that particular issue had always been a matter of concern for me and my fellow educators.)

The nation is again deciding who is crucial to society and who is not. We, the victims, our lives and our city, have already been placed on the back burners of importance. We are the backbones of society, but society was debating whether or not we were important enough to consider getting help to rebuild. We were being blamed because we were foolish enough to think that our unique and precious city was of any importance to this country. New Orleans was a great city. It can be great again, but only if the "powers that be" think we are worth it.

Washington was pre-occupied with investigations about who knew what, and when did they know it. Who did what, and when did they do it. Who told who what, and what did they do about it. Washington was so mired down in its own superiority that they forgot about the people they were supposed to represent. If they would have spent as much time and effort in helping the victims, many of us would be a little better off. We would know if we would be allowed to re-build in our old neighborhoods or if we would have to relocate. These important decisions were still being ignored because Washington wanted to find out who did something wrong, instead of finding out what to do right. AT that point we didn't care who was to blame! ROME WAS BURNING WHILE NERO FIDDLED!

Spring. So many were still struggling to make sense of their lives. I spent my time fixing what needed to be fixed when and how I could. I used the help I got where I needed it the most. My immediate family members and I helped each other as best we could. I was able to continue with my plans of living a new life of retirement while working on the damage to my house caused by flooding. Rita's life has changed because her two daughters were on the opposite sides of the world. Lauren and Kaylee were still in France and wanted to stay there. Alyssa went to Japan and married her boyfriend. They were not planning to return here for at least a couple of years. Nicolle comes to wash clothes and look for houses, but hasn't found anything yet. Cherie's life is almost back to normal. But rising prices have cut into everyone's budget, most of all, the price of gas. Why do the giant oil companies keep reporting profits but continue to raise our prices?

Conditions here on the north shore were slowly improving with little evidence that there was a hurricane. Most businesses have returned to their regular schedules. More people moved here from the New Orleans area, and there was more traffic. Blue roofs were slowly disappearing, damaged trees were being removed, and buildings were being repaired. A visitor would not be able to notice the severe damage we had. The new bright green leaves provide cover for the fallen trees that lay behind them. The blooming flowers divert our attention away from the remaining evidence of the aftermath of the storm. The feel of spring, the sights and sounds and warming days were a gentle reprieve from the cold winter and colorless aftermath of the storm. But the internal personal damage and displaced people could not be observed by an outsider. The New Orleans area wasn't recovering as easily. A visitor to that area would think that the storm had just hit. The destruction and emptiness was still there as in the other hard-hit areas. New Orleans, Chalmette, New Orleans East, and Lakeview looked the same as when I drove through months ago. A few houses looked like someone was trying to clean up, But mostly they looked empty. There were many "for sale" signs on most every street, and the real estate section of the newspaper listed plenty of transfers. People were buying houses "as is" at half their value before the hurricane. Many of the homes like my aunt and uncle's were abandoned because older established people used to live there and they had no one to help them with the clean-up. My aunt and uncle can never rebuild or return. They are too old. They remained in Texas near one of their children's homes and were still struggling about what to do. Their health is declining. My other aunt in Seattle has been diagnosed with Alzheimer's and doesn't even remember what happened. Even if someone in the family wanted to rebuild their homes for them, they would not be able to live there. Having no possessions left, they would be living in an empty house with none of the familiar things they saved for years and years. Besides, the fear of it happening again is too much for them to endure. The stress of relocating and starting a new life is also very difficult for them. No matter what they choose to do, the memories of the hurricane will haunt them. They will not live long enough to get over it. I had read in the death notices about an elderly couple, both just passed away within months of each other. They were evacuees living in South Carolina with relatives. Are their deaths going to be attributed to the hurricane, or considered just two more elderly people who died? Would they have lived longer and happier had they been able to live out their lives in their own homes?

Summer. The new hurricane season had arrived. People were already buying extra supplies just in case. Everyday the newspaper reported the progress of the rebuilding efforts and of the federal aid still on its way. Everyday there were also reports of people still homeless, still wanting to return but nothing to return to. Those that had their insurance money or other resources were frantically trying to finish their homes and businesses before the next hurricane strikes. In fact the sales of building products and home furnishings were so huge that the state actually had an increase in funds from taxes collected. But many people have still not made the decision of where to live, to go back or move away. They are haunted by the memories and threats of another hurricane. Maybe the younger people have time on their side. They might be able to recover from another disaster, or have time to start a new life, make new friends, new routines, and leave the past behind. Those in charge are still sending out conflicting messages. The politicians are practically begging people to return, while the scientists are saying that it is not safe, and they don't know if it ever will be. No matter how high they build the levees, they cannot know if there will be a bigger storm. This battle with Nature can never be won. No matter how resourceful we are, we cannot stop earthquakes, tsunamis, blizzards, volcanoes, tornadoes, or hurricanes. The best we can do is to protect ourselves from them. Or we can choose to live safely away from their threat of damage. Now is the best time to learn to work with Nature and not against Her. Let Her have Her coast back. There is an ongoing conflict about spending the billions of dollars it would take to reclaim the wetlands and build levees twice as high as before. But if our tax money is spent reclaiming the wetlands, who is to say that the next hurricane won't just wash it all away. Why isn't someone planning a better way for us to live instead. It would cost far less to rebuild neighborhoods further inland. Leave the coast for Nature, and we can go there for our recreation and enjoyment. I am reminded of the Gulf Coast area that used to be beautiful beaches, then gigantic hotels were build on the coast where we could not see the beaches anymore. The hotels became possessive of their beach area and would not let people who were not registered go on "their" beach. They convinced the people that the income would help improve the states infrastructure and education system. Money was used to improve highways and schools, but it could have been done differently. They didn't have to take away the beaches. Then the same thing happened with the casinos. Mississippi and Louisiana received millions of dollars promising the public to use that money to improve roads and schools and other things. Mississippi used the income to do most of that, but Louisiana did not. Louisiana's teachers and schools are still at the bottom of the list. Big business usually wins. Money and power rules again. Can we ever change that?

We need leaders that can understand the needs of the ordinary people as well as the big businesses. We need someone who will not be pressured and/or bought by those with power. Now in the Greater New Orleans area and in the entire Gulf South, we need someone with a vision of reality, someone who will tell the public the truth. Some areas will not be safe to live in. Period. Especially in an area that is below sea level. Especially New Orleans. It is over. It is time to find a new dream.

News reports are full of fraud, both within the government, FEMA, and even some ordinary people trying to capitalize on others' suffering. Money and resources were misused and some of that help is still sitting somewhere waiting to be doled out. Red tape, rules and regulations are keeping that help from reaching the people it was intended for. The government and private individuals are investigating and discovering corruption everywhere. Our fears of Nature's ability to devastate us is ever present, but our fears of what the people can do who are supposed to help us are intensified. Government agencies are dragging their feet and focusing on other things rather than deciding on what we need now and how to get it to us. It was bad enough to have to endure what Nature did to us, but I fear that the worst is yet to come. What Man is capable of doing is far more unbelievable.

EPILOG

New Orleans could become an illustrious world city. She could become a place that could set an example for the world as the most innovative place for diversity, acceptance, and ingenuity to take seed. She could become important for being the first city of the future, of a world that is that ever diminishing in size. She can teach the world how to be tolerant of every person, religion, and idea. She can attract the greatest minds, educators, builders, scientists, and religious leaders to develop a new way of thinking, a way that could change not only the South, but the entire world. But first, she would have to become independent. She would have to belong to everyone, and be answerable to no one.

Transportation should be one of the first priorities. A rail system like those in Japan and Europe needs to be built that takes people from Baton Rouge through New Orleans to the Gulf Coast, with plans to extend to other parts of the country. Rail systems, like those in Disneyworld, France, Japan, and most other large cities, would branch out from the main system to bring people from around Lake Pontchartrain and across the Mississippi River to and through the city. Traffic jams, parking problems and pollution would be lessened and more people would be able to get to the collages, universities and businesses in the heart of the city. The New Orleans area had one of the worst transportation systems. We could soon have the best. Politicians in Orleans Parish and the surrounding parishes would not work together to provide a decent system for the people. Because of their territorialism, the parishes suffered and the people suffered. Millions were spent on study after study that suggested how we could do it better, but nothing was ever done.

Education is another priority. But without the transportation to get the people to the schools, an excellent education system will not flourish.

There was an abundance and diversity of colleges and universities to attract students from all over the world. Each school has their own special offerings, but instead of being territorial, they should work together to provide the best educators to teach any and all subjects to everyone. They should also work together to provide the best education to the elementary, middle and high schools in the area. In fact, as an educator myself, I believe that the colleges and universities should have more control over the education system, independent of government control. Instead of having the poorest schools in the country, we would have the richest. Educators would decide how to educate children, not politicians. Politicians have always and are still proclaiming that education is their most important priority, yet when they get elected, suddenly they can't find the money to do what they promised. Money can be found for everything else, but not for teachers or schools.

Every child should be required to learn two or more languages from his/her first day in school not just here, but elsewhere in this country. Being able to communicate to the rest of the world is becoming more important each day. Even the poorest of immigrants can learn English. If they can learn two languages, why can't our children learn another language? It would be a huge benefit for everyone. We cannot continue to remain isolated in our own self-importance.

Before we can have great schools, we must have the best buildings to house them. All of the old school buildings should be torn down, not remodeled. They were in bad shape before Katrina, now they are more dangerous and dirty. New schools can be built first in the areas where the public is returning. Then new schools can be built in more areas as the public repopulates these areas. There will be some areas where no one will or should return. Why remodel a building where it will possibly never be used? It also costs less to rebuild than to remodel. There are many places in the Greater New Orleans area where people were tearing down old houses to put new ones up because it was more cost efficient to do so. The only objectors to this were the historic preservationists. The New Orleans schools are NOT historic buildings. If anyone thinks so, let them take over the buildings and do something else with them.

The arts should be returned. That was the first to be cut in a declining budget. Fine arts and music should be part of every child's education, especially here in a city that depends on the arts for its very survival. Art was everywhere, especially in the French Quarter. Jazz was everywhere. Without art and music, Mardi Gras would be nothing. The French Quarter and Bourbon Street would have had nothing of any interest to offer tourists.

If we want to educate children for the future, then it is necessary to teach them how to live in the future. If it requires the lengthening of the school day and year, then so be it. There is no reason why schools should not be open until 5:00 year round. An increasing number of children were enrolled in before and after school care and spent eight hours a day at school because there were more and more single working mothers, or both parents working. That time could be made more useful than babysitting. Parents send their children to summer camps, but if the schools were open year-round, the children could be spending their summers learning. Teach the languages and arts then. Teachers could be given a choice to work a six or eight hour day, or year-round, and be paid accordingly. Many work second jobs anyway because of the poor pay situation. This country found a way and the means to integrate computers into every classroom even when the teachers were not trained to use them. Why do they find it necessary to cut out the arts, not teach languages, but put greater emphasis on sports? The answer is painfully obvious!

After many earthquakes, engineers developed a way to build buildings to withstand the shaking of the earth. After major flooding problems, the Dutch developed a better system to keep the waters out. The Japanese built land in the water to build an airport that withstood earthquakes. Dams are built in rivers to hold back water from flooding towns and cities, and as a side benefit, ended up with recreational lakes. We can move mountains and dig tunnels through the earth and under water. We can send humans to the moon and back, and robots to explore other planets. We can clone almost any living thing. We can destroy complete forests to build new cities. We can cure diseases. We can dig a canal to connect one ocean to another. We can build ships as big as cities. We can do anything we set our minds to do. We can build a city over swamp land. We built a twenty-four mile long bridge across Lake Pontchartrain. Yet some people object to rebuilding New Orleans because of the cost. If the government cannot part with the money to rebuild, then let the world do it. Other countries are already contributing funds and others would like to. Let the world build the new city. Let the new city be governed by independent visionaries who see the promise and hope we can become.

We could have a new international airport, either by expanding the current one, or building an entirely new one on land that is now useless and dangerous for housing. Done correctly, similar to Japan's new airport, it can be built over swamp land between New Orleans and Baton Rouge, an area that would be assessable to both cities. I can already hear the environmentalists complaining. They have a good point, but where will the billions of people on earth live, work and play? I am not for destroying Nature to build cities, but if the best engineers and planners are involved, we can work together to do this right. I am also against trying to stop Nature from doing what She has to do. She has earthquakes, tsunamis, blizzards, volcanoes, hurricanes and droughts. We cannot stop these things, so we need to learn to work with Nature, not against Her. Nature destroyed our coast, not Man. Are we going to spend billions to restore the coast? Did we spend billions to restore areas damaged by volcanoes?

Nature restored Herself after the eruption of Mount St. Helens. Man was discouraged to live there after that event. People object when they are discouraged to live in a dangerous area because they have lived there before and want to return to their life-long homes, but the cost in lives is too high. We should legislate safety, because the cost of ignoring safety comes out of everyone's pocket. People want to know why they have to pay higher and higher insurance premiums because others choose to live in an area constantly and consistently in the path of Nature's destructive forces. We have the technology to improve our lives and protect ourselves, but we can only go so far. Should we build fire-proof homes so people can live near a volcano and not be harmed? Should we build water-proof floatable homes so people can live near tsunami and flood prone areas? Should we build igloo-type homes so people can live safely in blizzard areas? We can. Many engineers and architects have already done so. But should we? Those who fight for our freedom to live where we please forget that freedom is not free. Our freedoms are costly to all of us, and there will always be enormous decisions about where to draw the line between your freedom and mine. That decision has always been in the hands of politicians since the beginning of this country. Are they the best qualified to make those decisions for us? I think not. I believe it is time for a change in the way we govern ourselves. I believe we need a combination of people from all walks of life to run this or any country properly. We need people from the sciences, educators, health care personnel, economists, religious leaders (I do not support combining church and state, but they represent enormous amounts of the population), leaders from all races, etc. Some will argue that we already have that, but these representatives in Washington only have the power to suggest, not enact.

We here in the New Orleans area could do this. We could become a world-wide mini-experiment in a new kind of world government. We can do it and we should, if those in control would give up their control and give this a chance. What is the alternative? If we do not do something different, we will end up the same as before. I think the time and opportunity has come to take a chance, to show the world that there is a different way, one that even if it doesn't work, at least we can say that we tried something.

Engineers, planners and builders were innovative enough to build casinos on barges, why can't we build our schools on barges. The perfect place to start is the University of New Orleans (UNO), since it is right near Lake Pontchartrain. Because of the flood waters, they were severely damaged. If we build new classrooms on barge-like structures, the classrooms would float when it floods again, and it probably will. We could build enormous ferries similar to those in Seattle, that would bring students from across the lake in the Mandeville and Slidell areas to New Orleans. While transporting students and faculty, classes could be held on the ferries, along with coffee shops and small stores. They would not need to build more parking spaces, and the schools would be more accessible to more of the public.

New structures could be built in the same way, on barges. If we can build bridges like the many we already have, miles and miles of raised highways, why can't we build everything on raised bridge-like structures. We can use the waters of the lake and river, let them flow where they will, under and around the buildings to transport people and goods to the city and through the city. With a good transportation system, homes and apartments don't need to be so close to the city. There is so much available land in the parishes all around the Greater New Orleans area that new residential districts could be built in those areas that did not flood. The flooded areas could be transformed into parks and green spaces so that when we have another disaster, people would not be in the way, lives and possessions would not be lost. Many scientists predict that the entire area will be under water anyway in the not too distant future because Global Warming will melt the glaciers and ice caps and the sea level will rise significantly. Whether they are correct or not, I would not want to take a chance and invest my future based on that information. It would seem that the logical thing to do would be to relocate the residential areas if nothing else. For the environmentalists, we would just be trading one green space for another. Nature has spoken by reclaiming the area south and east of New Orleans, we should listen.

There are people around the world who are innovative, talented and visionary. There are people who know that there is a better way, but are hampered by closed minded people who have the control and don't want to give up their "cushy" positions or "rock the boat." Visionaries have always been ostracized by government and religious leaders. But often history has proven that the visionaries were right. The world is not flat, and our dreams should not be either.

Many new kinds of problems arose because of Katrina and because of the attacks of 9-11. We felt secure and complacent before. The events of 9-11 taught us not to trust People. Katrina taught us not to trust Nature. Now we all live in fear. Part of the fear is that People and Nature took something away from us. Beside lives and property, they took away our sheltered existence. But it also gave us a new perspective on what is important. Our nation came together after 9-11 and we all felt a new closeness with each other. But the feeling faded away with the memories. Then Katrina happened and those feelings came back. Once again, people of this nation and the world felt compassion and offered help. Once again, those feelings are fading. How soon we forget!

I don't have answers. I can't say what is right or wrong. No one can. Each of us has to believe what is right for us, but I do know that we need to stop criticizing each other for not believing in the same things that we each hold dear. This is why individuals fight with each other. This is why we have wars. This is why people are killed, because they believe that they are right and others are wrong. There is no right way. Ours is not the better way, only a different way. If we don't start accepting each other's differences and their right to believe what they want, if we don't learn tolerance, then we will continue on until we destroy ourselves and there will be nothing left to fight for, and no one left to fight it.

We must do something different now, while the memories are fresh. Something must change, and we are the only ones who can initiate that change. Each one of us, alone, but together.

On August 29, 2005, we had a nightmare. Now we have the unique opportunity to change that nightmare into a vision.

EULOGY

Time expired

Fate transpired

Lives mired

Emotions wired

People tired

Memories acquired

Dreams sired

Visions re-fired

Hope required